卓越家具

FABULOUS FURNITURE

深圳市艺力文化发展有限公司 编

大连理工大学出版社

图书在版编目(CIP)数据

卓越家具：英文、汉文 / 深圳市艺力文化发展有限公司编. — 大连：大连理工大学出版社, 2011.8
ISBN 978-7-5611-6327-6

Ⅰ. ①卓… Ⅱ. ①深… Ⅲ. ①家具—设计—作品集—世界 Ⅳ. ①TS664.01

中国版本图书馆CIP数据核字（2011）第134354号

出版发行：大连理工大学出版社
（地址：大连市软件园路80号　邮编：116023）
印　　刷：利丰雅高印刷（深圳）有限公司
幅面尺寸：245mm × 290mm
印　　张：17
插　　页：4
出版时间：2011年8月第 1 版
印刷时间：2011年8月第 1 次印刷
责任编辑：初　蕾
特约编辑：陈　静　胥群燕
责任校对：仲　仁
装帧设计：王　佳

ISBN 978-7-5611-6327-6
定　　价：298.00元

电 话：0411-84708842
传 真：0411-84701466
邮 购：0411-84703636
E-mail：designbook@yahoo.cn
URL：http://www.dutp.cn

如有质量问题请联系出版中心：（0411）84709246　84709043

FABULOUS
FURNITURE

PREFACE

Design in the interplay between the unicum and serial production

Stefan Heiliger Design Studio

The unicum is an individual product that reflects the needs and ideas of one person and is usually driven by emotions and spontaneous inspirations. The private use of the unicum by its designer and his friends completes its subjective direction, as well as its artistic impetus that also articulates social criticism and irony.

The serial production has clearly a broader reference to the society. It reflects the requests and needs of as many users as possible and tries to express social streams and trends with which the future owners of the products can identify. The acceptance of the market throughout a longer period of time is a precondition for economical action.

While the unicum can give food for thought, the serial product must create more reasonable and longer valid expressions of social reality.

However, both forms of design are unified by the communication between the user and the product. Each piece of furniture and each accessory tells something about itself, as well as about the lifestyle ideas of its addressees.

Therefore, the focus on a certain awareness of life or on a particular lifestyle always means a restriction with regard to the quantity of the serial production.

Nevertheless, the user respectively the owner also communicates through the product: On the one hand he navigates the way he wants to be perceived and with which period of time he wants to be linked. On the other hand he defines both his level of luxury and his social status.

But there also exist inner modes of operations, because lifestyle products stimulate different behaviour patterns. Consequently, the user feels either interpreted by a soft curved relax lounger, or by an ostentatious sofa design.

Further, sharp-edged linear forms stand for precision and intellect. Thus, symbolism is an important and influential term.

However, it always applies that processes of social change function as a catalyst for design. Therefore, modified social and economic factors demand reconsiderations and new interpretations of all characteristics of the product. In order to understand a design or to visually receive it, one must know the individual cultural background of the product.

The surrounding area of the user also offers him modes of identification, depending on his individual phase of life. Thus, the communication between user and product becomes a dense and complex relation. Hence, its modes of operations are used by the advertisement in order to reach the market. Nevertheless, the market and the literature is flooded by a broad spectrum of designs. Thus, it becomes obvious that the user product relation cannot so easily be exploited for the market. In consequence, the instinct for trends and future lifestyles and needs will always overlay the rationality of planning and marketing.

Frankfurt am Main, July 2011
Prof. Stefan Heiliger

CONTENTS

FABULOUS FURNITURE

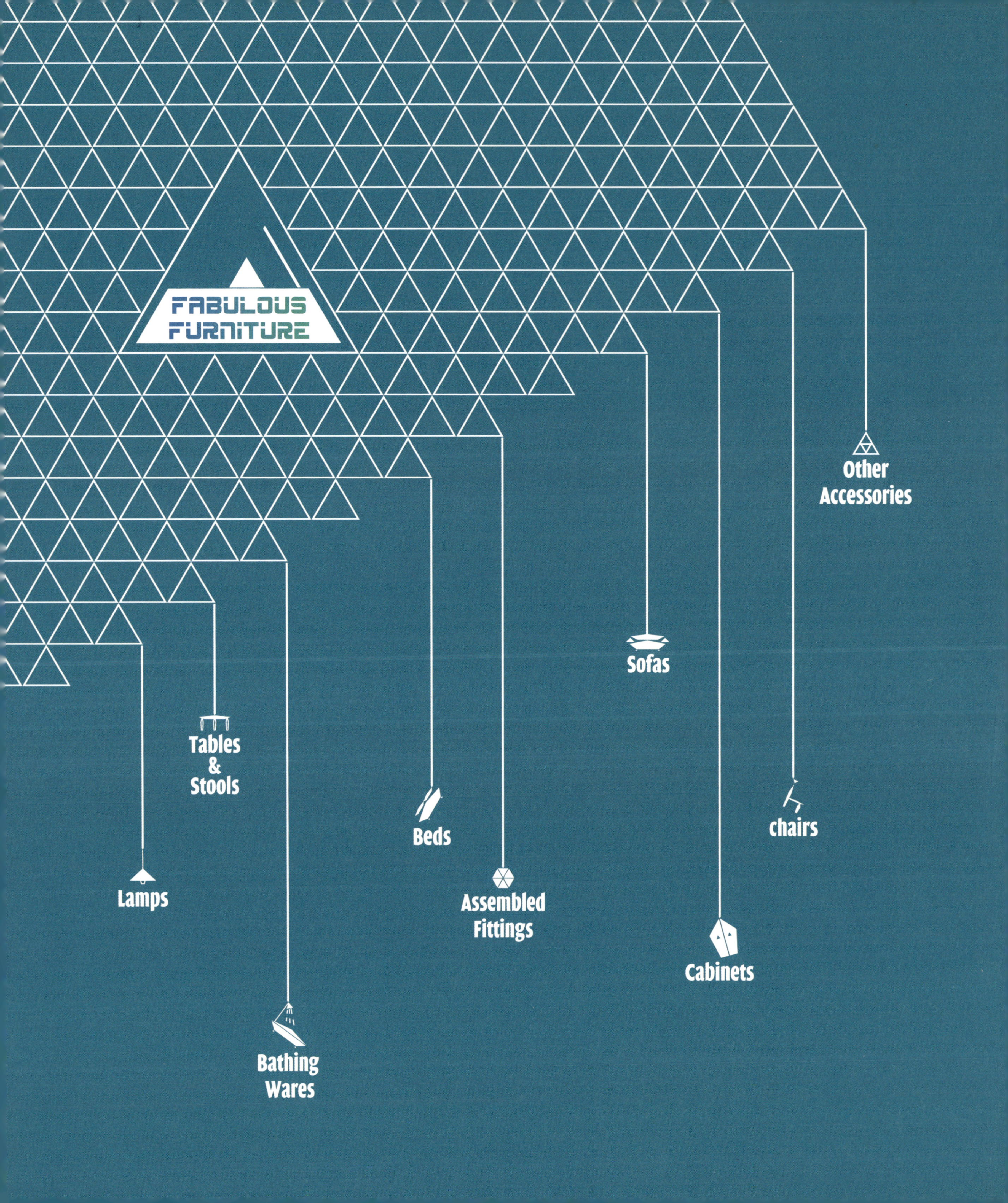
FABULOUS FURNITURE
Other Accessories
Sofas
Tables & Stools
chairs
Beds
Lamps
Assembled Fittings
Cabinets
Bathing Wares

SICIS

SICIS entered into the Big Apple, with the official opening of the flagship Store in SOHO, New York in July of 2006.

The creation of SICIS North America coordinates the distribution of SICIS collections through its growing network...following the SICIS philosophy to create an advanced and well-organized distribution system, up to the satisfaction of every request from all over the world.

LUCKY CHARM

Design Agency: SICIS

Enchanting adornment for hair left loose 4 leaves I give you, fortune and pardon I accuse.

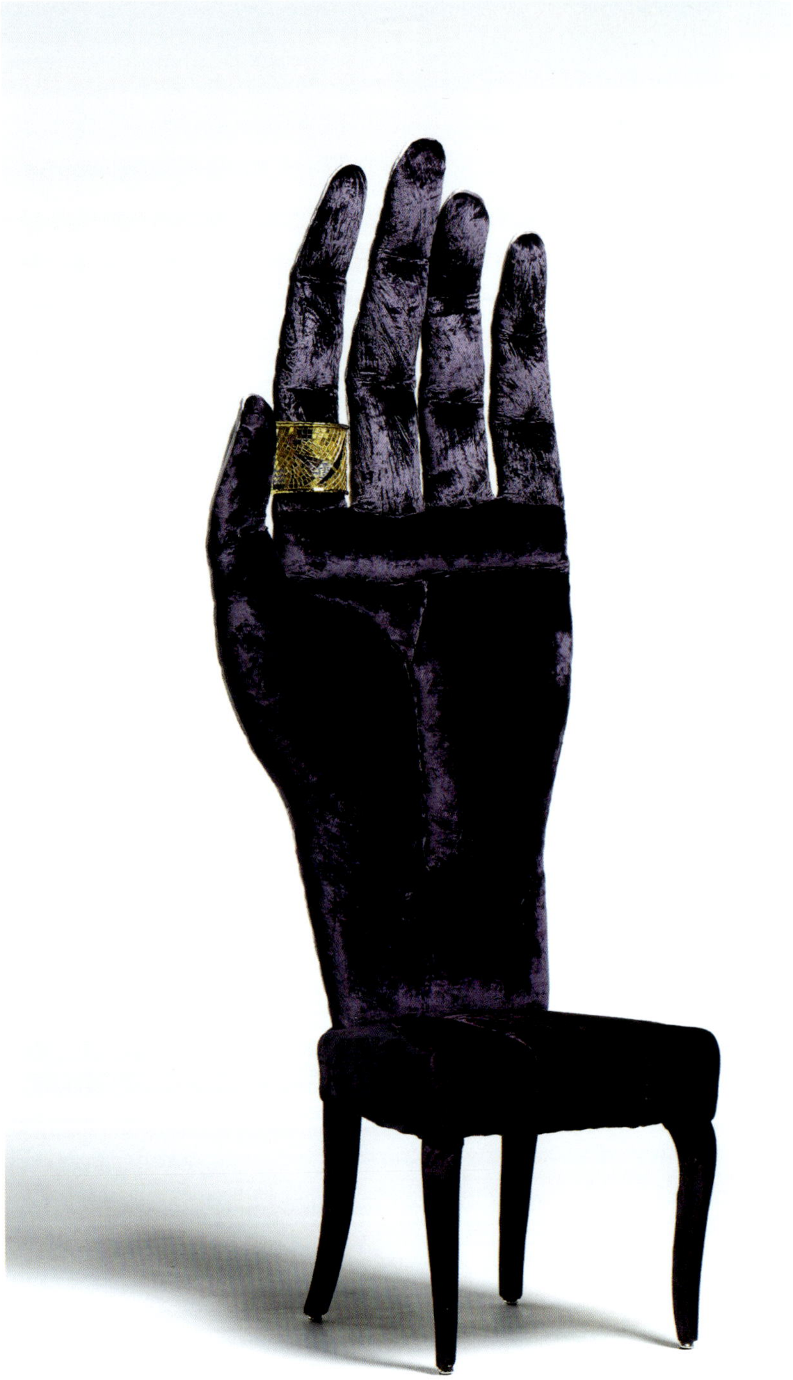

The big promise

Design Agency: SICIS

With hands, if you like to release a memory, distant and nostalgic.

The black moon

Design Agency: SICIS

Perfume of moon, magic and spicy, laid out, in folds, at the peak of luxury.

Moi et la Rose

Design Agency: SICIS

Like buds that open in verse, the breath of crystal petal flowers set in place.

Le Rouge et le Noir

Design Agency: SICIS

Red sap of life amidst a blend of hands, amidst weaves, the lines of destinies and lands.

Nina Bruun

Nina Bruun is a 26-year-old Copenhagen based designer. She works in the fields of furniture, spatial and graphic design. Her main focus is furniture design—good handcraft and graphical expression is a must. Nina Bruun is educated from The Danish School of Design and is specialized in furniture and spatial design. In the spring of 2010 Nina Bruun won 1st prize with her Nest chair in a big competition named SYDFORM for designers from Sweden and Denmark (60 participating designers). And was together with the FOLD chair, exhibited from March to September four different places in Sweden and Denmark.

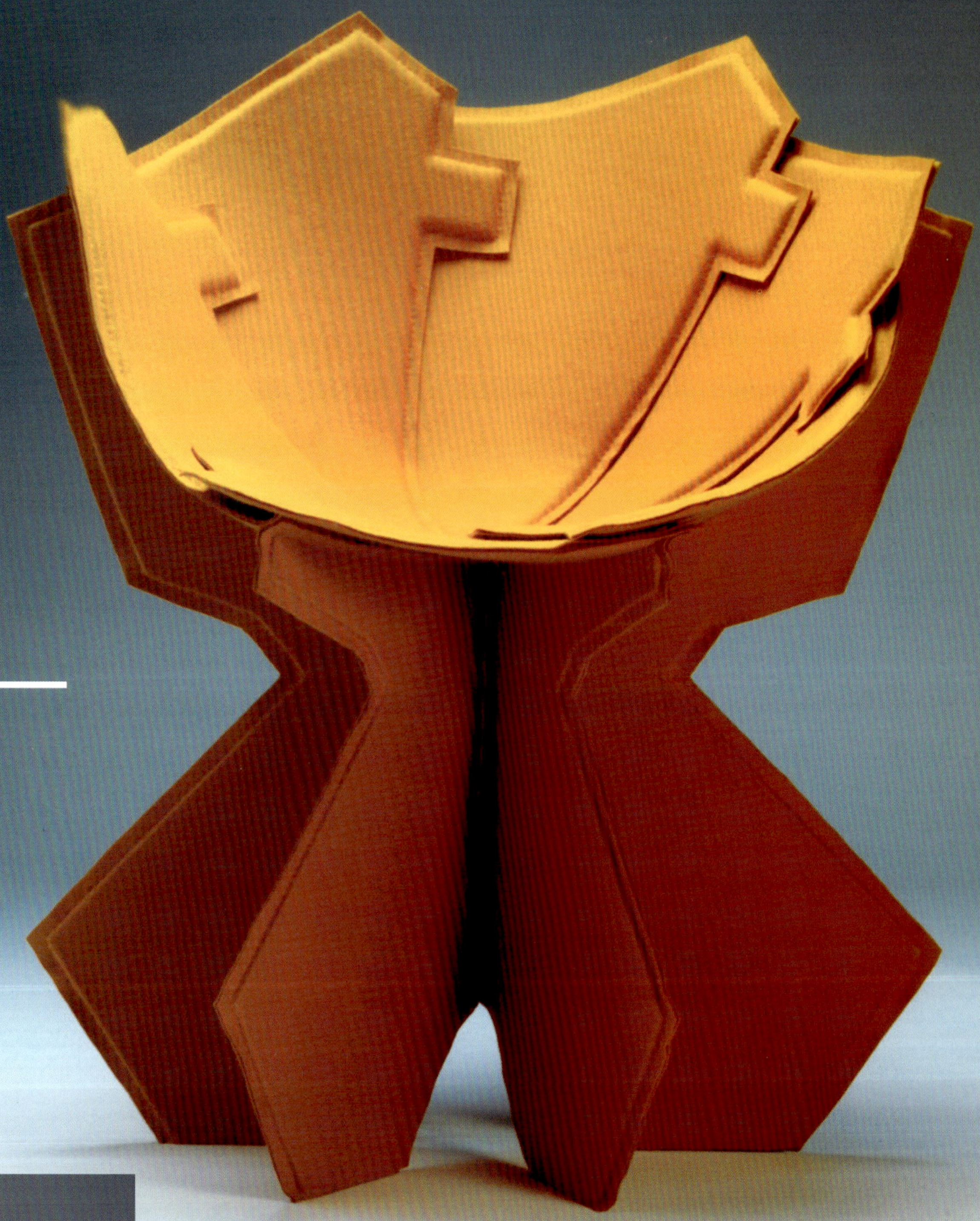

Fold

Designer: Nina Bruun

Nina Bruun wanted the chair to be foldable and to have references to Scandinavian and Japanese design. The shell consists of six pieces of hard plastic, which is upholstered with woolen felt. All seams on the chair are hand sewn, which took her 105 hours. The seams are visible to create a more industrially expression and to create sharp edges which give the chair a lighter expression. The look and function of the chair is inspired by paper folding—especially the Japanese origami folding technique.

NEST CHAIR

Designer: Nina Bruun

Influenced by the spring and by nature, this lounge chair is inspired by a bird's nest. Nina Bruun's idea was to create a graphical chaos which still felt like a unified whole. To create a 'nest feeling' Nina played with birch strips in different thicknesses and hurled them around the supporting structure. It was natural for me that the chair had a soft center to create references to the bird's nest, that is typically lined with down, to form a confident frame for eggs and chicks.

Fehling & Peiz

Yvonne Fehling and Jennie Peiz are based in Karlsruhe and Munich, Germany.
In 2006, they founded their design studio and the brand 'Kraud'.
Yvonne Fehling
Assistant professor at the University of Karlsruhe.
Working experiences in the Netherlands and China.
Jennie Peiz
Working experiences in the U.K., India, Australia and the Netherlands.

Still lives

Design Agency: Kraud
Designer: Yvonne Fehling and Jennie Peiz
Materials: smooth leather, skin color
Photography: Frederik Busch

Objects for domestic space.
Fitment, toy, luxury good, sculpture. Stuffed toy, gym equipment. Projection area.

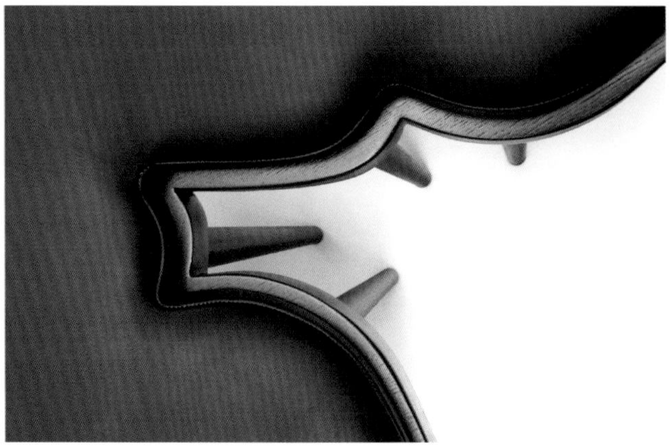

HOCKERBANK

Photography: Frederik Busch

STUHLHOCKERBANK

Design Agency: Kraud
Designer: Yvonne Fehling and Jennie Peiz
Materials: solid oak, oiled finish
Photography: Horst Bernhard, Fehling & Peiz

'Stuhlhockerbank' is a series of seating for public spaces that approach users and viewers in an extraordinary way. The dividing lines between different types of furniture (chair, stool, bench) are eliminated as the three types blend. The results are immobile constellations.

SUPER-LIGHT

Photography: Daniel Schludi, Piero Glina, Johannes Tolk

Daphna Isaacs & Laurens Manders

The Eindhoven based twosome creates a broad range of design projects by uniting their individual qualities and interests. Both graduated at the Design Academy Eindhoven (2006 and 2008) they now run their own design company.
With their colorful way of working Daphna and Laurens transform everyday objects into characteristic products. They research tradition, material, composition and translate it into colorful contemporary forms.
Next to their research approach everyday thoughts, reactions on human 'cultural' behavior and evoking emotion are themes that pop up in their work. Shaping by hand and intuitively making design choices give their work an authentic appearance.

Vault

Design Agency: Daphna Isaacs & Laurens Manders

This vault's appearance is inspired by the mineral pyrite, also known as fool's gold. Because of its resemblance to gold pyrite is often mistaken to be highly valuable being actually almost worthless.
This now resembles my vision on value. You can steal the vault and/or it's 'valuable' contents but you will find out that it's of no value to you...

TAFELSTUKKEN

Design Agency: Daphna Isaacs & Laurens Manders

The five 'complete' lighting objects shine upon themselves as well as the objects placed in them. The lamps are open for your own interpretation. For example, a 'Tafelstuk' can be used as a container to highlight your everyday objects; keys, fruits or a magazine. All the pieces are produced in Daphna and Laurens' Eindhoven based workshop.

Ontwerpduo

A dreamer and a mathematician, this young Dutch design duo complement each other well. The wild fantasy is created, and is then surreptitiously grounded by the mathlete; It's a philosophy that seems to keep things from going a bit over eccentric, and tends to generate an alluring, yet graspable charm.

Designers Nathan Wierink and Tineke Beunders—both graduated cum laude from the Design Academy in 2008—just couldn't wait to get their imaginative paws immersed into the design marketplace. While studying, this duo nurtured the fledgling design studio. The pair make up a formidable force, and business acumen aside, they admit they just couldn't delay the fun any longer.

Garden of Eden

Design Agency: Ontwerpduo
Material: coated steal
Photography: Lisa Klappe

Garden of Eden is a set of furniture for the garden. The hanging chair can be placed in a tree or veranda, like a big birdcage. It gives you the feeling to sit in your own space. At the same time you can have contact with the surrounding because of the open structure.

The chairs and table have a laser cut pattern from different items you find in the garden. The same pattern is embroidered at the pillows.

The furniture can be placed in the garden as a piece of jewellery, and when you want to use it, simply add the pillows.

ontwerpduo

What it is, it isn't

Design Agency: Ontwerpduo
Dimensions: instrument 140cm x 70cm x 70 cm, cupboard 200cm x 95cm x 50 cm

Objects are how we perceive them to be. Our minds make things what they are and for us this is reality. Think of a branch hanging in the water. We see, with the refraction of light, the branch under the waterline in a different angle. But we know that in reality the shape of this branch is different. We learned how to see the world.

High loft

Design Agency: Ontwerpduo

Ontwerpduo was asked to design the interior of a high loft in The Hague. The space is four meters high, and with that ontwerpduo designed a huge shelf-wall with adjustable color accents. The living room and the bedroom was completely designed in this concept. The design, the making off and the result of the room was broadcasted by RTL4, in the program 'Eigen huis en tuin'.

Studio Geenen

Studio Geenen is a start-up design studio working on progressive techno-logical projects. Meaning that the projects seek solutions by implementing new technologies in a logical and natural way, whilst preventing from thinking and acting in the most conventional or common ways, since these are by no means always the best ways to go. Until recently the resulting products remained limited to furniture and lighting, whereas at the moment the studio is also working on advanced online and virtual projects.

GRID SHELF

Design Agency: Studio Geenen

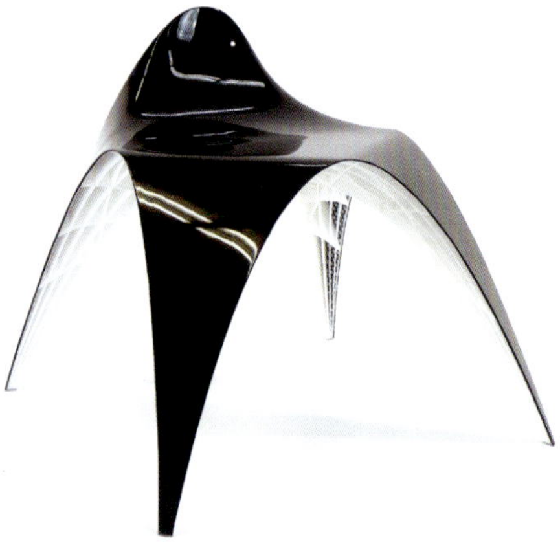

GAUDI CHAIR

Design Agency: Studio Geenen

The Gaudi Chair is the follow-up of the Gaudi Stool which was created in 2009. It was designed using the same method as Antoni Gaudi, who made models of hanging chains, which upside-down showed him the strongest shape for his churches. Additionally, to be able to determine the structure of the chair's backrest, a software script was used.

The script was based on three steps: First, the distribution of forces across the surface of the chair. Secondly, the direction of forces defined the direction of the ribs. Finally, the amount of force specified the height of a rib.

DYNEEMA CHAIR

Design Agency: Studio Geenen

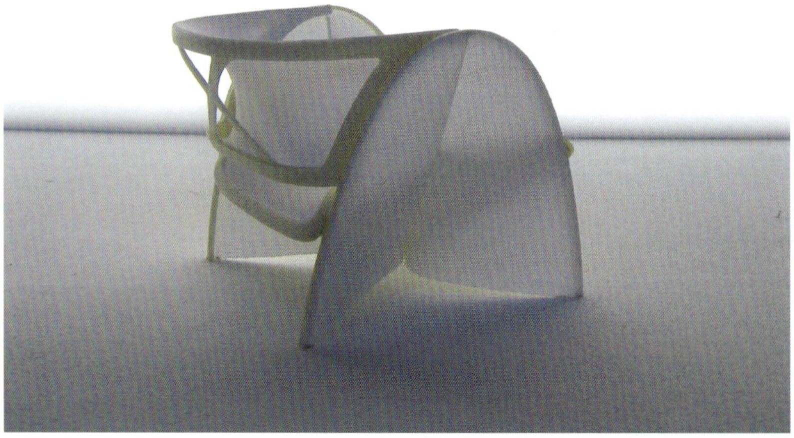

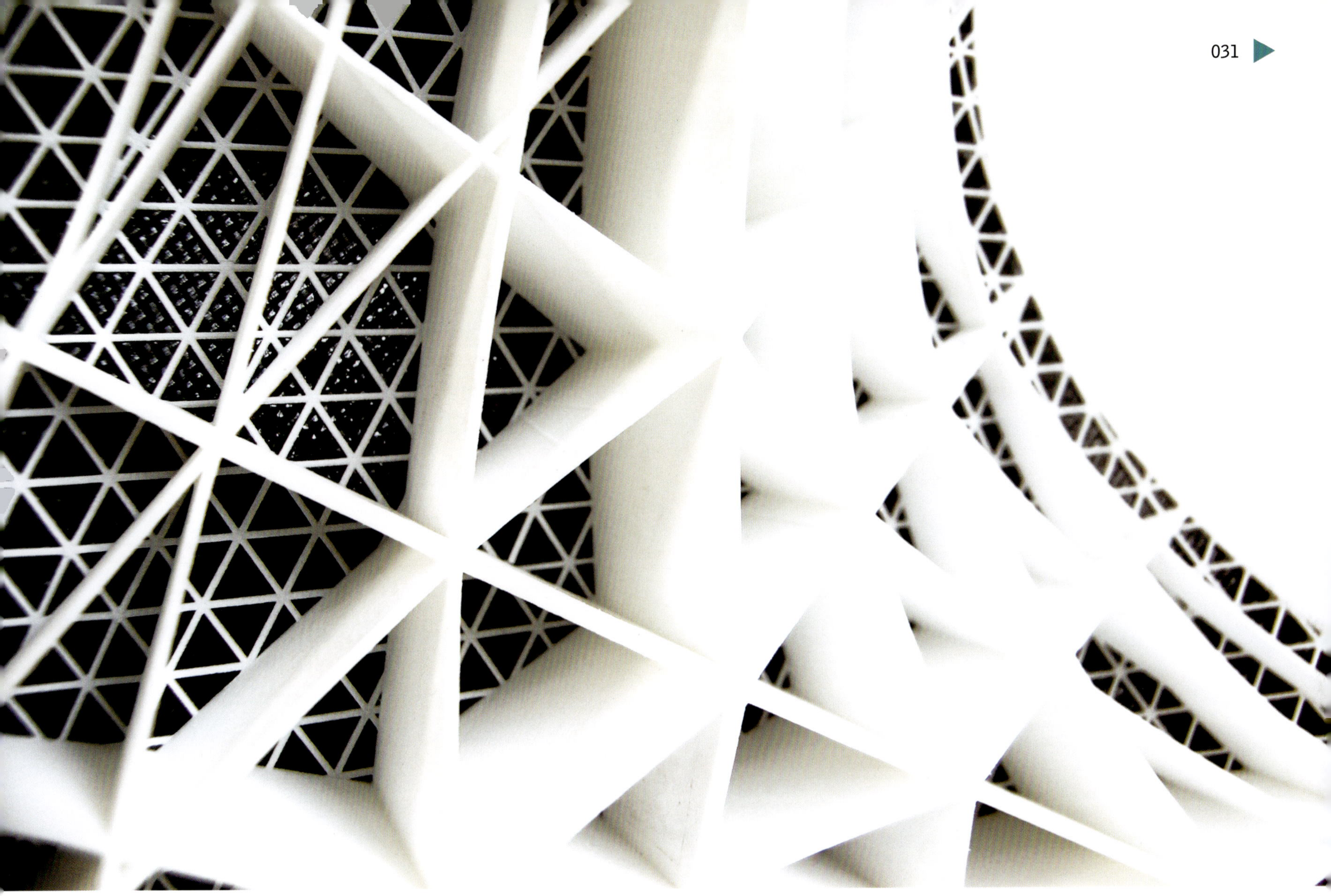

GAUDI STOOL

Design Agency: Studio Geenen

The shape of the Gaudi Stool was created in the same way that Antoni Gaudi designed the structure of his churches, by making a model of hanging chains, so letting gravity determine the strongest and most logical shape for withstanding forces.

Materials and techniques: The stool consists of a shell made out of carbon-fiber, and a substructure which was made using 3D-printing techniques. The product weights exactly 1kg.

Michaël Bihain

The kind of precise research that results in the obviousness of the lines obtained: a symbiosis of contemporary lifestyle and collective memory. Michaël Bihain is the designer that won't terminate the creative process. Instead, he seeks to hand over the creative process to the client, or owner. The objects are meant to work as solitudes—or by repetition create a new unity. Bihain hands over the 'relay baton'—like the awarded, unforgettable, ladder looking bookshelf Libri for Swedese.

Having trained both as a butcher and carpenter, Belgian Michaël Bihain studied Interior Design at the St Luc Institute in Liège and Hull University. Following his studies he opened his studio, currently working on the development of furniture design and accessories either from his office in Bruxelles or from his workshop in Louveigné, both in Belgium. Michaël Bihain is teaching at the St Luc Institute in Liège and has been awarded several prizes. The Design Museum in Stochkolm and Seoul has some of his creations in its permanent collection.

Mosquito

Designer: Michaël Bihain
Manufacter: WildSpirit
Dimensions: 80cm x 57cm x 47cm
Material: curved plywood

The 'Mosquito' chair is the result of a thought about humanism and the autonomy of furniture. It is a first step towards the 'no furniture'. The result takes the form of a living being: an autonomous, obvious chair which gladly volunteers to play the role of a comfortable seat. The 'Mosquito' chair is made from a leaf of waisted wood. It may be piled up in two ways: pragmatically or graphically.

Iris

Designer: Michaël Bihain
Dimensions: 35cm x 91cm
Material: lacquered steel

The coffee table 'Iris', made from a single sheet of steel, comes in white or black lacquered steel. 'Iris' can be piled up and is suitable for indoor and outdoor use.

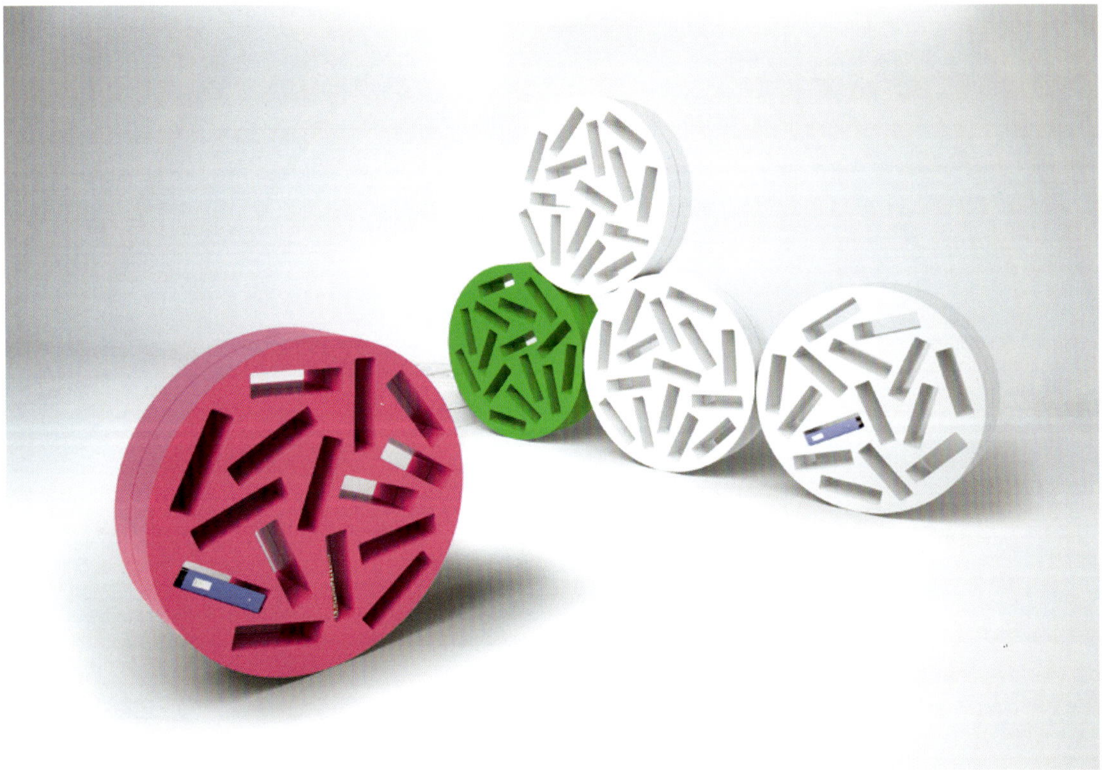

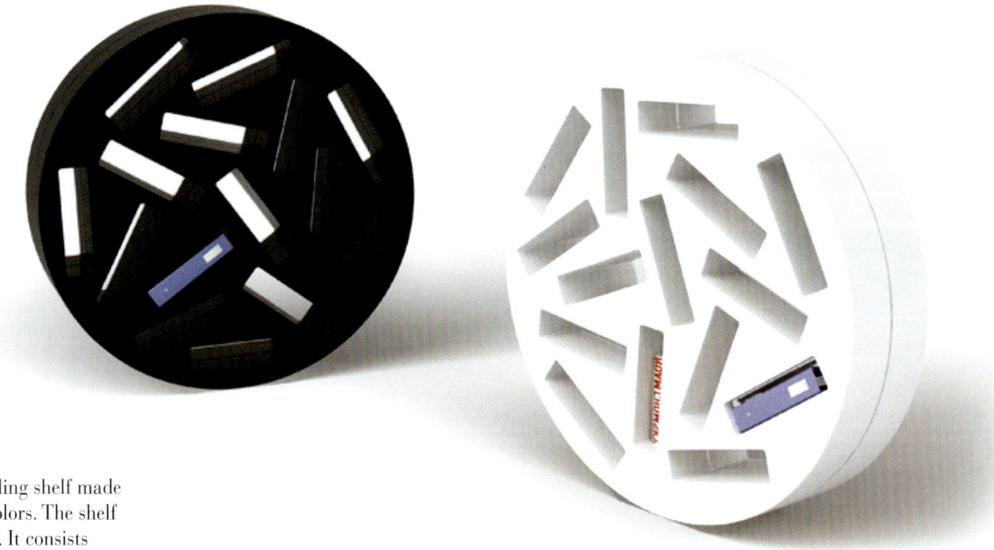

PATATRAS

Designer: Michaël Bihain
Dimensions: 122cm x 29cm
Material: expanded polypropylene

Patatras is a new kind of books storage. It presents itself the rolling shelf made of expanded polypropylene and available in a range of bright colors. The shelf could be interpreted as a wheel of knowledge or a round library. It consists of many single cells for one book each. With Patatras you should be selective about the books.

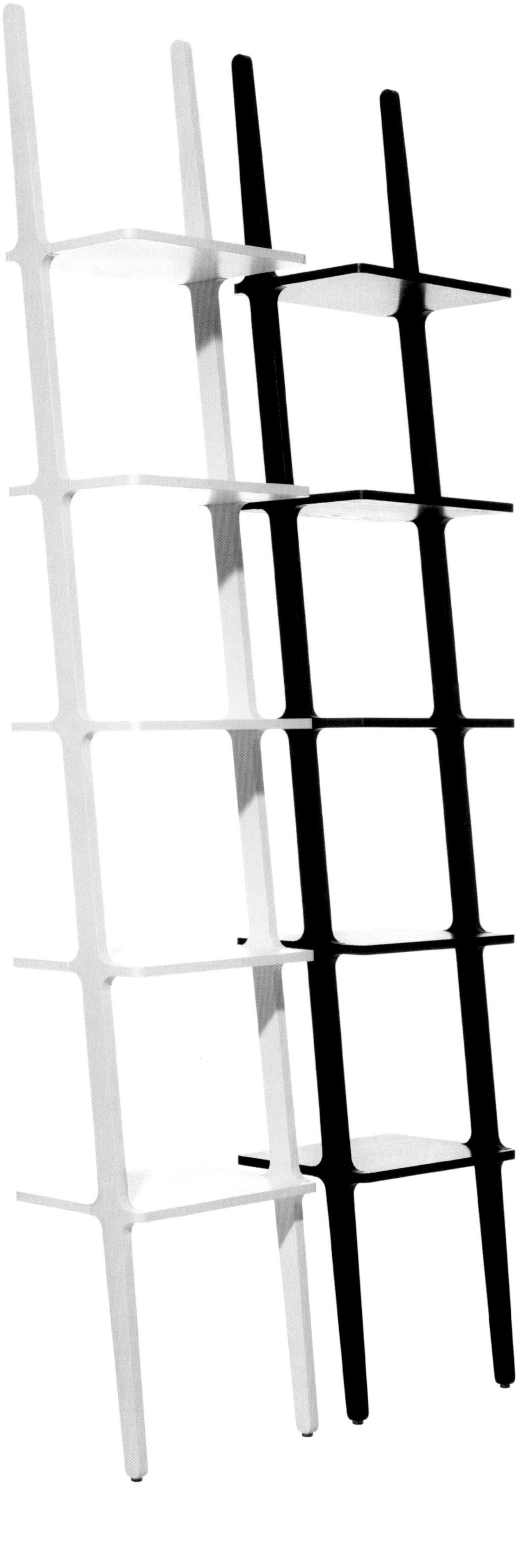

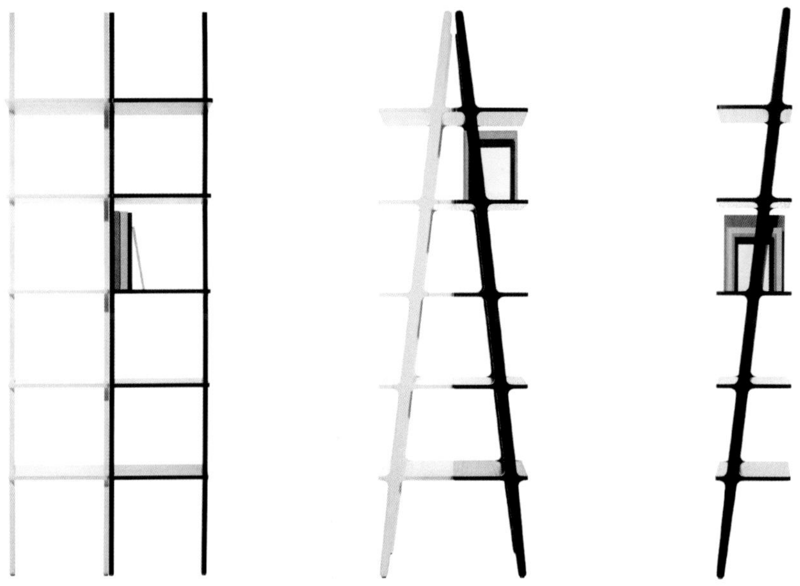

Libri

Designer: Michaël Bihain
Manufacturer: Swedese
Dimensions: 227cm x 38cm x 29cm
Photography: Mikael Strinnhed

Libri shelf in white or black lacquered ash wood can be used as a single unit attached to the wall. It can also be combined with additional Libri shelves assembled in width along the wall or right out from the wall as a room divider or even back to back to become a free standing shelving system. The legs are equipped with adjustability in height for uneven floors, and they can be adjusted after the shelf/shelves has/have been attached to the wall or to each other.

Florian Gross

Florian Gross lives and works in Tübingen. Born in 1982, he completed his training as a furniture carpenter in 2001 and began studying interior design before successfully graduating in 2009. Since then he has worked as an interior and product designer.

KONNEX

Designer: Florian Gross

Konnex is aimed at people who enjoy constantly redesigning and organising their surroundings. And with this innovative new slot-in system, exactly that is possible. A basic set, comprising three modules, can be individually arranged to form a shelf unit. In addition, different numbers of basic sets can be added to form a unique shelf space: the comb-like slits are inserted into the side walls of another cube to form a large shelving system. The way you decide to piece them together is variable depending on your preference and the occasion and can therefore be adapted to every space and every situation.

Yellow Diva

Furniture design, by definition, ought to be an anthropocentric discipline; Yellow Diva places the body at the heart of its design philosophy. The intervention of the human being imbues the sculptural object with purpose and adds a dynamic component. Yellow Diva products are classic and enduring. The studio encourages responsible consumerism and advocates simplicity of purpose. Recognizing the intrinsic importance of good design Yellow Diva believes that carefully selected products add value and meaning to our surroundings.

M SERIES

Design Agency: Yellow Diva

Taking inspiration from Eastern and Western traditions the M Series embodies relaxation and well being. Upholstered in 100% eco worsted wool, with solid oak legs and oversized felt buttons, the M Series lends a whimsical twist to Yellow Diva's collection.

FANDECK

Design Agency: Yellow Diva

Dimensions: Flat: 410mm x 100mm x 20mm/ Assembled: 200mm x 600mm

This award-winning, flat pack shade is inspired by the efficiency and elegance of natural flora and fauna. Simply fan out the plywood petals of the shade and hang from your own pendant cord. Once suspended, gravity generates the final three-dimensional form.

Les M

Designers Céline Merhand and Anaïs Morel have worked together under the Les M label for three years. Graduating from the Rennes Regional School of Fine Art, their degree shows were sponsored by the VIA and displayed in the gallery of the same name in 2007. Nominated at the City of Paris 'Grand Prix de la Création' in 2010, they exhibited their creations at the Maison et Objet trade show in September, and joined forces with the young Luxembourg design company Superette to create 'Cocon', 'an armchair with a new typology, the irrefutable union of a seat and a duvet leading to the most comforting seat'! (Duende PR)

'COCON' ARMCHAIR

Design Agency: Les M
Dimensions: 55cm x 100cm x 80 cm
Materials: removable and washable duvet, zip fastening
Photography: eric chenal/ blitz agency

'Cocon' is an armchair with a new typology, the irrefutable union of a seat and a duvet leading to the most comforting seat for reading, rest, and viewing on the market. An obvious item, inspired by a close observation of our daily routine, 'Cocon' fulfils maximum snuggling up pleasure for long TV sessions.

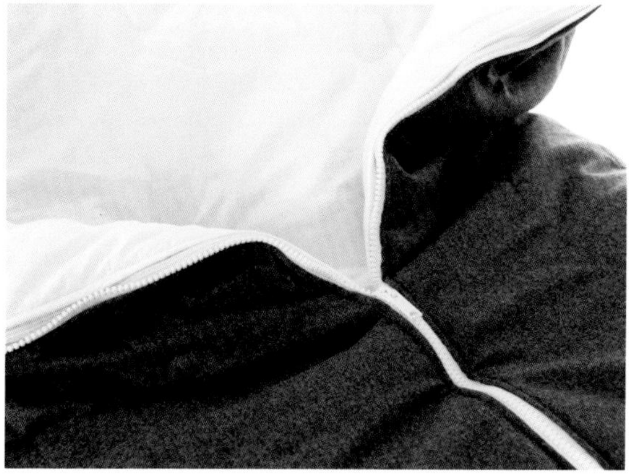

Collerette

Design Agency: Les M
Dimensions: 45cm x 48.5 cm/ Blanket size: 145cm x180 cm
Material: fabrics

Collerette is a footrest and a comfortable seat. Once its collar unrolled, the user can wrap himself up in the integrated blanket.

Jorge Pensi

Jorge Pensi: Architect and Industrial Designer, was born in Buenos Aires, Argentina. After starting his business labor in Argentina, he moved to Spain, where got the Spanish Nationality. He was established in Barcelona, taking part of Berenguer Group, collaborating with Alberto Lievore until 1984. In 1984, he founded his own's Studio specialized in furniture and lighting design, visual image of his product and assembling for several kinds of events. The team of Design of Jorge Pensi is integrated by Diego Slemenson, Constanze Schütz and Roman Proubasta. Toni Casares is the General Manager and carries the Management out of the Studio. Besides, his wife Carme Casares collaborates in the graphic design projects. He has worked for the following companies: Akaba, Amat, Andreu World, B.Lux, Perobell, Santa & Cole, Cassina, Driade, Frighetto, Thonet, Knoll International, Kron.

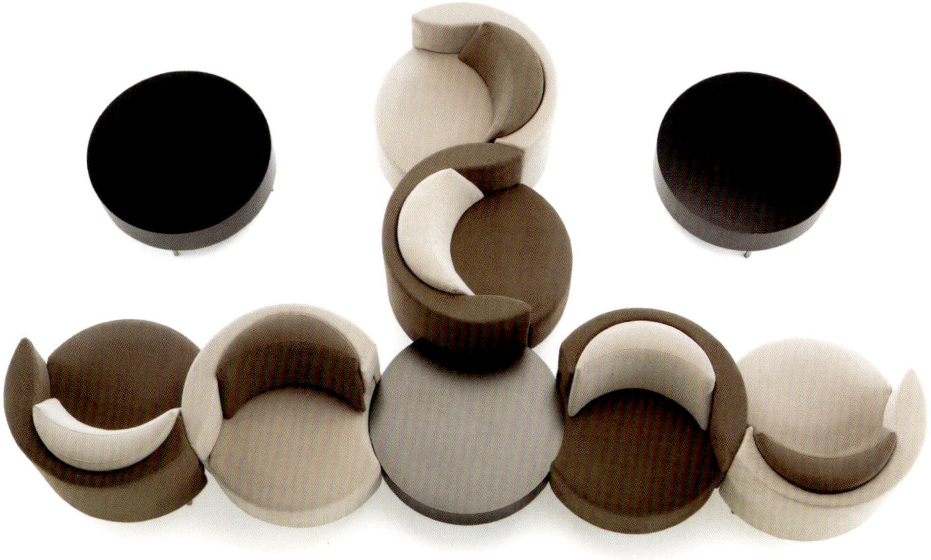

PLAY

Design Agency: Frajumar
Designer: Jorge Pensi

As its name shows, the armchair 'play' lets several compositions, creating atmospheres both in formal and informal in the architectural environments. Recreational design but content: in individual armchair, it starts from a circular base, which joins together other armchairs, poufs and tables, getting unheard-of collections.

Gosi

Designer: Jorge Pensi

The offer of the Gosi sofa is simple: it offers absolute comfort. Its high back lets rest the head on soft cushions. Its incline arms give the possibility of sitting down sideward. Visually, it's opened to the user. It is an object defined by the synthesis between austere and light lines and comfortable shapes.

Rick Lee

San Francisco based industrial designer Rick Lee finds inspiration in everyday life experience. Lee's theoretical approach to design is not unlike an actor's tactics. Working with a wide range of materials and resources, he prefers not to use any one particular material on a consistent basis. Lee considers 'California Cuisine' an apt metaphor for his art—a mix of seemingly incongruent ingredients carefully combined to create something inspiring and exceptional.

DUNE SEATING

Designer: Rick Lee
Materials: seating: polyurethane foam, cotton fabric covering, rug: hand knotted from Nepal in wool and silk.
Dimensions: Seating: 198cm x 107cm x 67cm/ Rug: 198cm x 366cm

An homage to the desert's natural elements, the multifunctional dune seating/sleeper rug unit is a brilliant cornucopia of butterflies, geckos, poppies and birds. With the rug's spellbinding serpent design gliding across the soft curves of the seating unit, it serves as a sofa or becomes a lounge sleeper for two. Making the ultimate statement of style, it can also be transformed into a back-to-back seating unit for intimate soirees or casual gatherings.

LACE TABLE

Designer: Rick Lee
Dimensions: 44.5 cm x 44.5 cm x 53 cm

Delicate and feminine, the table's intricate cut-out pattern is inspired by ancient Chinese lattice window designs. Highly refined and architectural, this piece leaves a soft yet modern touch to the viewer's eye. Superior design details include hand-finished miter joint construction and a forty-five degree edging.

CNC TABLE GROUP (4 SEASONS)

Designer: Rick Lee
Material: sustainable 1/2 inch thick plyboo in amber color.

Celebrating the cyclical nature of life, the CNC table group is a harmonious symphony of the fleeting seasons. Fusing CNC technology with an artisanal touch, the result is a perfect amalgamation of nature and modern design. Each serving a different function, they express their own unique persona, from the wintry coffee table with barren tree branch patterns to summer's fiery pedestal table.

Mask love seat

Designer: Rick Lee
Dimensions: 152cm x 76cm x 76cm
Material: cast aluminum with high polished finish.

Originally designed to celebrate the San Francisco Opera Ball opening. Fusing the language of art and design with whimsical touch, the Mask Love Seat is both a sculpture and a fluid economic seating.

MOJAVE CHAIR

Designer: Rick Lee
Dimensions: 66cm x 83cm x 68cm

Named after the iconic desert, this signature piece imitates the fluidity of vast sand dunes. Perfect for meditative reflection, the chair features a deep and ergonomic lounge seat.

SUBLIME SEATING UNIT

Designer: Rick Lee
Materials: fabric over polyurethane foam, wood frame and down feather filled throw pillow. As shown in Knoll textile. Available with or without casters.

In the three-section Sublime seating group, originally designed to celebrate the opening of Sublime American Design furniture gallery in New York, Rick Lee has developed a system in which the unlikely geometry of the circle has been negotiated in such a way that all three components can fit together in numerous configurations.

Mardi gras lamp

Designer: Rick Lee
Materials: stainless steel, LED rope light and floor foot switch.

A vibrant lamp based on the concept of a festive celebration, the Mardi Gras emits a bold glow from loops of LED rope lights which the shade is the light source. The Mardi Gras is as much an art piece as a functional lamp.

Swiss knife seating unit

Designer: Rick Lee
Dimensions: 279cm x 76cm x 66cm
Materials: fabric over polyurethane foam, wood frame and aluminum swivel mechanism. As shown in knoll textile.

Based on the iconic army knife, Lee's Swiss Knife seating unit carefully treads the line between form and function. Utilizing a blade-like aesthetic, lower sections on each side of the couch swivel 360 degrees creating myriad possibilities for positions and seating arrangements. The Knife is customizable and can seat up to five people when closed and 8 people when fully opened. The Swiss Knife's innovative form and striking red fabric create both an artistic conversation piece and a functional couch.

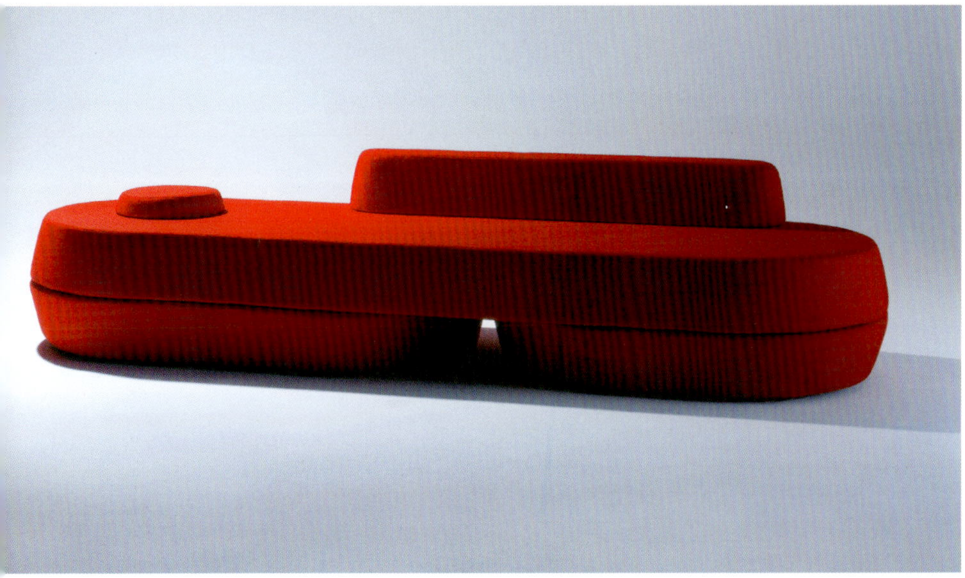

TARGA ITALIA

During its ten years' activity TARGA ITALIA has developed a research centered on proposals of furnishings and ambience for living and working.

Its activity carefully considers the stimulus coming from the design and art worlds as well as from the evolution of living. Along this course we have used the production mechanisms in a peculiar way : we started from our experience as designers. A way of producing where the search for rigour and quality does not refrain from going along new routes, from looking for innovatory technologies and materials.

SCARLETT

Design Agency: TARGA ITALIA

Armchair and pouf supported for interior or exterior use by a tubular metal frame and covered by a stretchy multilayer load bearing fabric with sheets of different thickness varying in coincidence with selected pressure points. The construction characteristics give to this seat special effects of lightness, sculptural quality and chromaticity. The seat covers are available in one color: black or white, or two colors such as red and orange, red and violet, white and orange, green and purple.

LONGUE

Design Agency: TARGA ITALIA

It is thought as a bookcase made of long 'segments'. By developing an innovative concept, it shows itself as a strong 'sign' whose proportions, out of ordinary measures, connote, through one or more elements, the wall of the living room or of the office so significantly that a simple definition of it as a bookcase is no sufficient to describe it. Because of these proportions, the result is that the entire wall becomes part of the composition. Internally a series of accessories can be inserted: boxes and baskets, writing-shelves, computer work station facilities. These accessories can be rotated or translated so allowing a great personalization and a wide functionality.

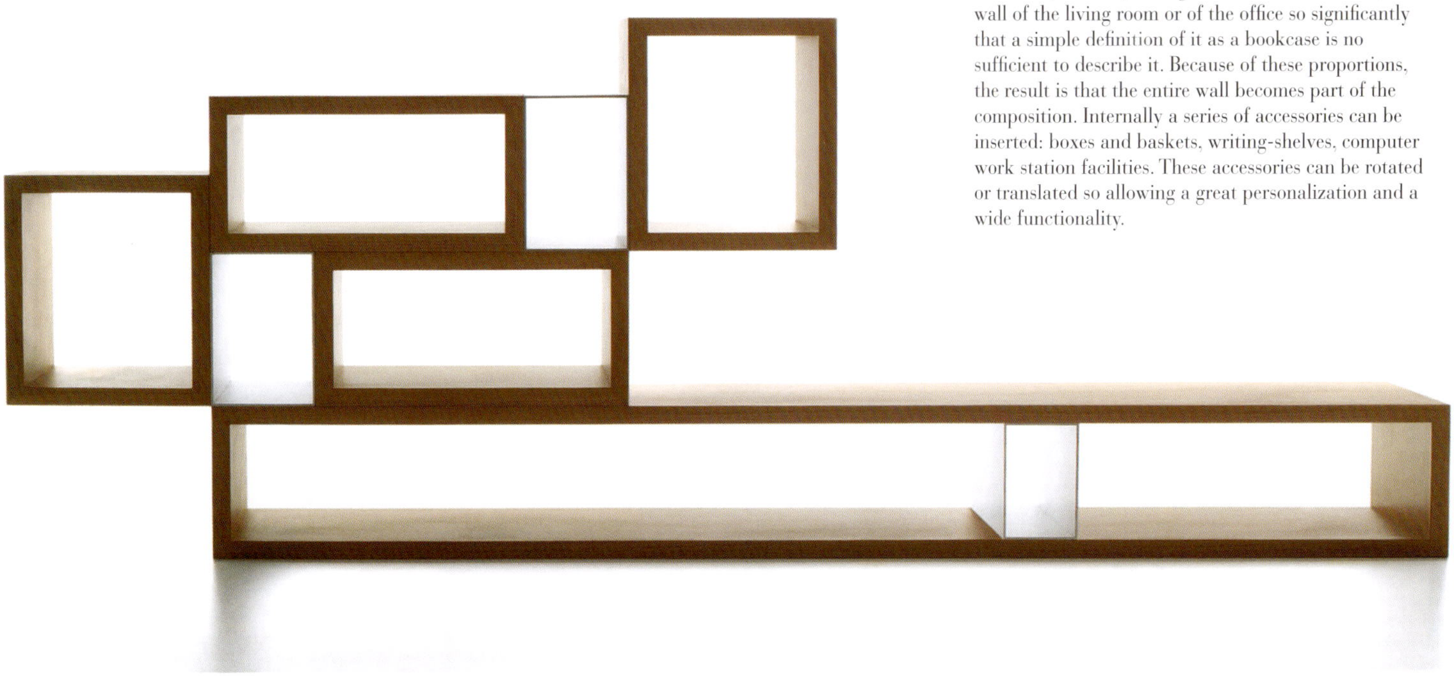

MALMÖ CHAIR

Design Agency: TARGA ITALIA

MALMÖ Chair is freely inspired by the Scandinavian design tradition. Seat and backrest are in curved plywood while the frame is in solid oak: a traditional technique still suitable to manufacture chairs with topical shapes.

浦东大道
PUDONG AVE
大连路隧道
DALIAN RD TUNNEL

ANITA SOFA

Design Agency: TARGA ITALIA

Anita Sofa is esthetically characterized by its low and modern lines which stress its horizontal dimension design. This characteristic does not reduce the seat utter comfort.

The armrest, positioned at a height complanate to the seat, enhances its formal characteristics: wide and protruding it becomes an extra space for sitting or serving. Its cushions, all of the same dimension, can be used as arm or backrest at pleasure, they offer the possibility to combine different colors and materials as they are not mere accessories but help to strongly characterize this model.

SALINA

Design Agency: TARGA ITALIA

This table is inspired by the shapes of nature. It doesn't follow the formal and functional pattern of the tables with regular geometry. Its sinuous shapes enable an easier relation between the object and the interior around it as well as a better use the man can do of it. Three are the available versions: small, medium and big. The top finishes are the following: wood in several iridescent veneers and lacquered in different colors with anti-scratch treatment. Metal legs can be chromed or lacquered same colors as the top.

Eduardo Climent

Eduardo Climent starts his business adventure in the middle of 2002, creating the firm ECDISSENY-ESTUDIO. He was hired by companies of design and develop of products in the furniture and lighting area, but his energy and tenacity lead him to extend his business vision starting his work in the interior decoration. In March of 2005 he got the first and second prize in the Design Contest of the Murcia Region organized by CETEM and with the collaboration of the company Frajumar. His design vision is large.

Nomi

Designer: Eduardo Climent

The project consists of divan of metallic structure, made in stainless and upholstered cylinders together with the same structure.
In his use, it has being used materials at the back zone and endorsement that allow the flexion backwards, according to the exerted pressure and its recovery once the user let to use it, obtaining an adaptation of the product for any person, independently of its weight and height.

Noga Berman

Noga Berman, a 1983-born young Israeli designer, is a recent graduate from the Bezalel Academy of Arts and Design, Jerusalem. Noga grew up in a rural community in Israel, and is a daughter of two goldsmiths. Her years of living and studying in Jerusalem exposed her to the colorful and enriching environment of the city. She says that for as long as she can remember, she was interested in hand-crafting objects from different materials. Shaping the sensory and emotional user experience is her primary guide in the design process, and in the attempt to create a meaningful interaction between the user and the object.

Tunnels

Designer: Noga Berman

Noga Berman created a human-scale object that can be used for lying down, sitting and storage. The technique of wrapping elastic bands creates tunnels, a perspective space, and peek holes. Thus, the user can choose his personal way of using the furniture. The technology used for making the objects is that of bending metal tubes and wood etching. Elastic bands are wrapped around these and closed with velcro.

Meritalia Spa

Meritalia was founded in 1987 and established itself in production of interior design characterized by an original mix of research and design. The company shows today to believe more than ever in the added value that Italian design can bring into the world: the main collection of the brand comes out from prestigious cooperations and, during time, is set with products that play at the border between art and design.

Nubola

Design Agency: Meritalia Spa
Designer: Gaetano Pesce

Nubola is a solid, comfortable and soft sofa which can be easily recognized by dimension and functionality. It is fraught of big colored buttons. The sofa seems to be created from a cloud. Its own solid and steady form of a common sofa changes, moulds, bellies out with the same texture of the air which is prisoner of various textile coverings.

Montanara

Design Agency: Meritalia Spa
Designer: Gaetano Pesce

Padding in polyurethane waterlilly. Upholstered in fabric, 100% cotton, digitally printed. Structure in steel and wood.

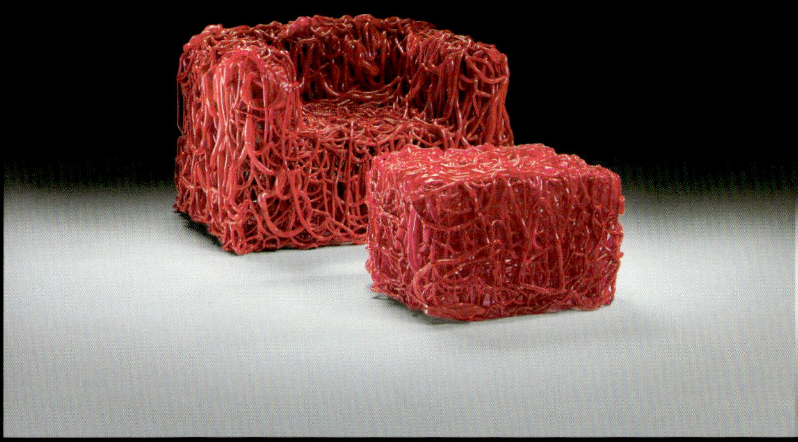

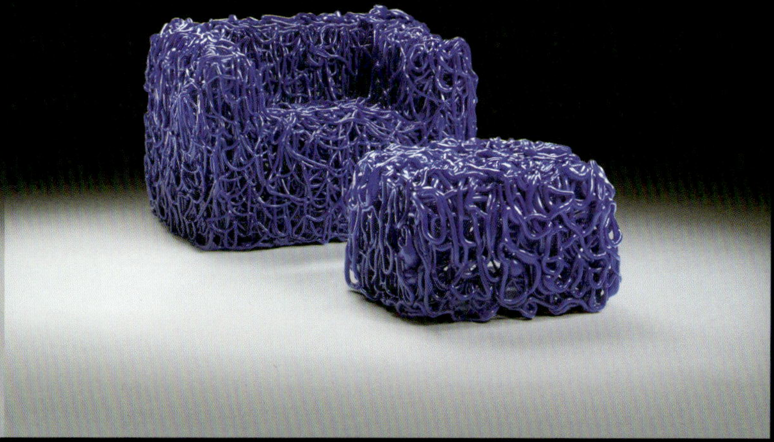

Senza fine

Design Agency: Meritalia Spa
Designer: Gaetano Pesce

This experimental series of seats is originated from the continuous extrusion of a silicone curb randomly disposed on a mould until the person who applies it considers satisfactory the form reached.
This process, of course, gives rise to unique pieces which, for the way of their execution, are unrepeatable. In reason for the strength of silicone to the elements, this series 'Senza Fine' can be placed both in internal and external spaces and from, this latter ability, comes the idea to expose these 'pieces' on the sidewalks of Via Durini in front of the showroom Meritalia.

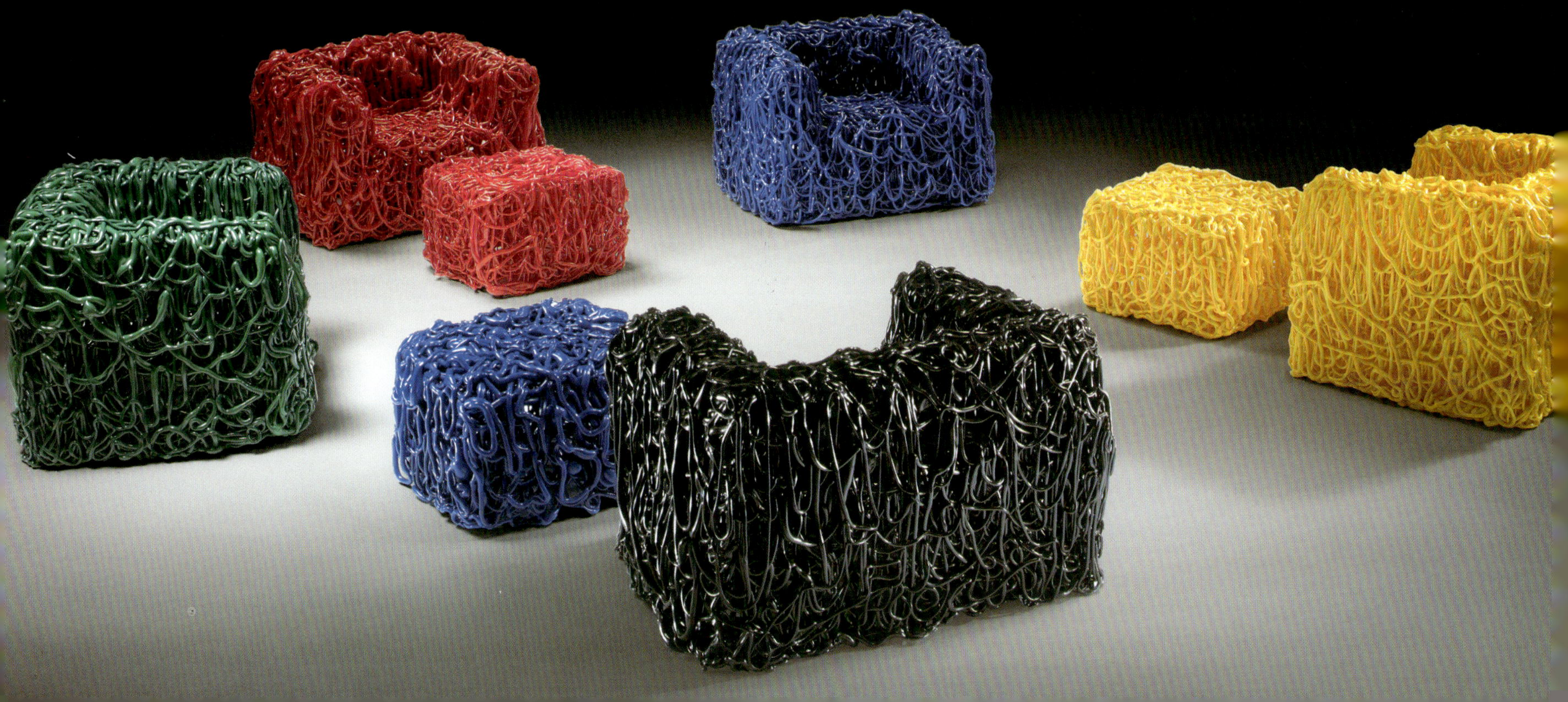

Stefan Heiliger Design Studio

Stefan Heiliger is one of the most prominent product designers known for his recliners, easy chairs and couches. He has designed strategic ergonomic, social and comfort transitions between 'sitting-lying' and 'lying-sitting' more than almost any other designer.

In his work diverse influences come together: first, familial heritage, his father Bernhard Heiliger's sculptural sensibilities; second, experiences gained during his studies at the famous Ulm School of Design, and during his time with Wilhelm Wagenfeld; and, last but not least, professional experience as a designer for Daimler-Benz.

With this experience it seems evident that Heiliger is as much an emotional functionalist as he is a sculpturally thinking pragmatist.

PARABOLICA

Design Agency: Stefan Heiliger Design Studio
Manufacturer: Leolux, Netherlands

QUESTION MARK

Design Agency: Stefan Heiliger Design Studio
Manufacturer: Tonon, Italy

The flowing form connects the seat, back and foot with a continuous movement, allowing the floating and free-swinging seat. The design goal was to achieve the greatest possible reduction in the chair configuration.

One for two

Design Agency: Stefan Heiliger Design Studio
Manufacturer: Tonon, Italy

Line

Design Agency: Stefan Heiliger Design Studio
Manufacturer: Bonaldo, Italy

The parallel guide principal serves as the inspiration for his piece. Through joints within the closed surface, both sitting and lying are enabled in a range of possible configurations. The hollow space within is transformed into function with the addition of a table element.

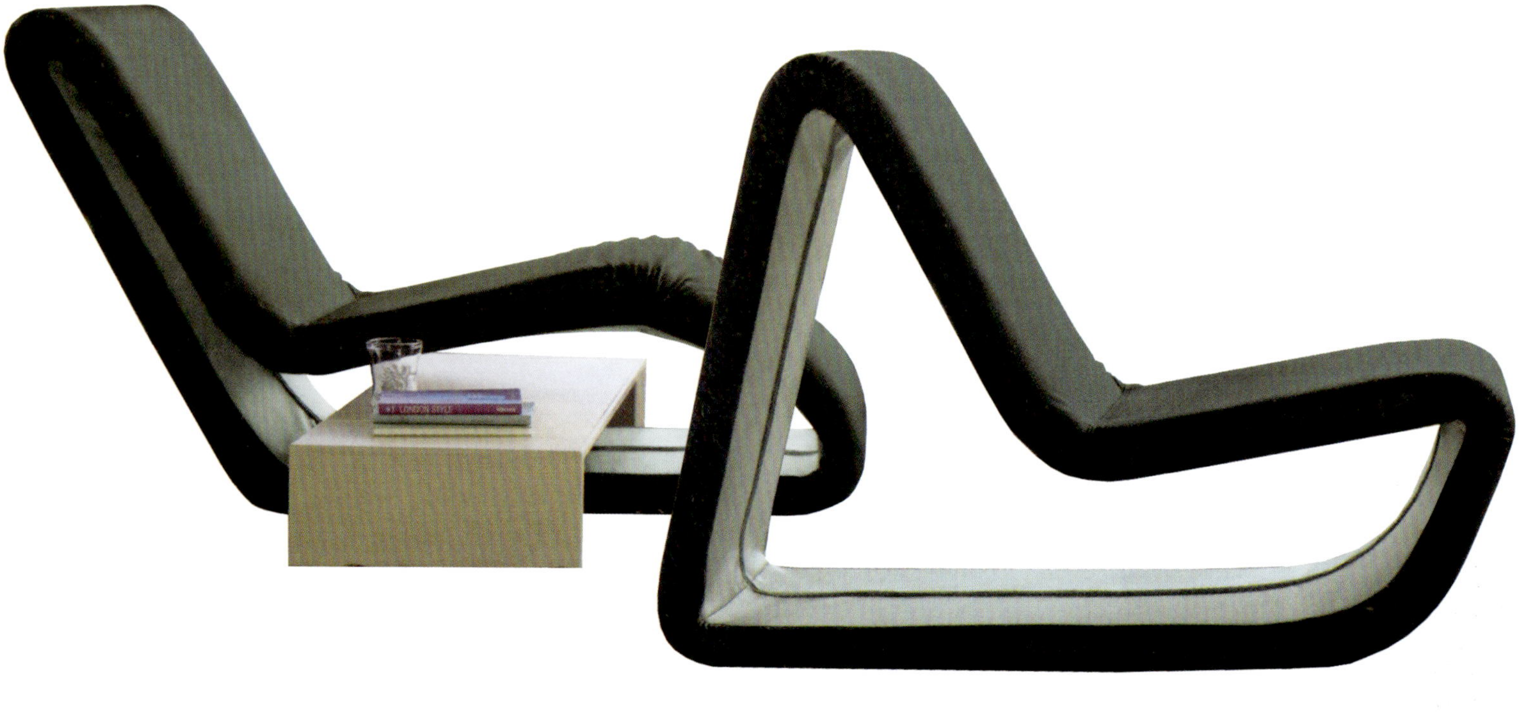

Morena

Design Agency: Stefan Heiliger Design Studio
Manufacturer: Leolux, Netherlands

The idea for Morena was that two or three identical or mirrored sofa elements can be made into each other or next to each other and so they can form variable landscapes.
The key message is the oblique flank, which oscillates from positive to negative, and so always provides the right docking area.

Orbit

Design Agency: Stefan Heiliger Design Studio
Manufacturer: TF-Polstermöbel, Germany

The purpose of this design is the fact that you can make the chair smaller. In this position the armchair is fully usable, however, it offers optimum comfort if you swing out the armrest and so you get a hemispherical hollow, which stimulates the user to many relaxed positions.

Koji Collection

Koji Collection is a Miami, Florida based furniture company committed to providing high style, ultra modern furniture and accessories made from natural, easily renewable materials. Here at Koji Collection we feel that 'going Green' is not just the latest trend, but is a moral decision to be part of the solution instead of the problem. Their goal is to provide a collection that modern living, forward thinking, eco-conscious people can be proud of.

GUI LIN

Design Agency: Koji Collection
Dimensions: 185cm x 80cm x110cm
Materials: white rattan, reclaimed wood, satin cushions

A futuristic vision in rattan! The Gui Lin's flowing lines and elegant presence are sure to add an artistic flavor to any ensemble. This stunning couch (and matching chair) is only one of the many Koji Collection pieces pushing modern organic design to new realms. Thick strands of meticulously inlaid rattan compose the silhouette of this lovely lady. Like any good diva, she manages to be glamorous without being over the top.

Ring

Design Agency: Koji Collection
Dimensions: 75cm x180cm x145cm
Materials: woven liana base, inlaid rattan canopy, canvas cushions

A rattan masterpiece! The Ring is Koji Collection's star piece and hands down winner when it comes to popularity. EVERYONE loves the Ring! Even children are immediately drawn to its naturally appealing aesthetic. This universal symbol of unity is sure to capture the imagination of any admirer. The Ring is also available for outdoor specific applications. The versatile 'Ring' is sure to spice up any space.

Shell

Design Agency: Koji Collection
Dimensions: 125cm x 117cm x 88cm
Materials: rattan, artificial leather

Purely organic in appearance, the Shell's flowing lines create a naturally appealing silhouette. Its deeply sunken seating position encourages relaxation and ease. Tightly woven liana vine is painstakingly wrapped around a wooden frame. Beautifully upholstered with pleated leather, this stylish accent chair provides a timeless design that will remain relevant for years to come.

Ripple

Design Agency: Koji Collection
Dimentions: 105cm x 185cm x 125cm
Materials: rattan, satin cushions

The Ripple is a crowd pleaser! You will be hard pressed to find someone who isn't left slack jawed at the first sight of the Ripple. The Ripple has a bold presence in any room, augmented by stunning craftsmanship and intricate detail, all topped off with a generous helping of billowy satin cushions.

BIRDY

Design Agency: Koji Collection
Dimensions: 134cm x 121cm x 127cm
Materials: woven rattan frame with PU leather, suede or black & white cow skin upholstery.

Inspired by the humming bird, the Birdy is yet another fine example of how Koji Collection pieces capture the harmony between exquisite artistry and precision craftsmanship. This fun piece quickly draws the attention of any passer-by. The deep sweeping lines of the Birdy's meticulously woven rattan frame converge into sharp angles, resulting in one of the most beautiful and unique accent chairs available. Tweet !

Living Divani

Perfect, harmonious proportions and a touch of understated luxury are the hallmarks of Living Divani, an enterprising, dynamic family-run business that has built its name on upholstered furniture.
Since the early 1970s, the company has blazed its own trail by establishing an intensive exchange with the contemporary design world, becoming a standard-setter in the field. Over the years, the company has put together a comprehensive range of sofas, armchairs, beds, and decorative accessories, through steady creative development and a production process that has gradually led to the creation of a complete living room collection.

FAMILY CHAIR

Design Agency: Living Divani

Collection of chairs with a remarkable magnified optical effect, for light objects, practical in their harmony.

Pasticca

Design Agency: Living Divani

Modular seating system that characterizes all settings with a carefree elegance.Spherical elements combined together to create an infinite number of compositions.

CHROMATIC SOFA

Design Agency: Living Divani

Light lines and precise shapes are the key features of the Piero Lissoni Chromatic sofa, with its visible metal or wood structure that is subtly, playfully, pared down. The metal serves both to define the outline and to confer a sense of lightness.

BUKVA

Design Agency: Living Divani

Victor Vasilev, a young Bulgarian architect living in Milan, has designed for Living Divani the modular bookcase Bukva, with its strong visual impact, steel structure and painted fronts.

Studio Floris Wubben

The designs of 'Studio Floris Wubben' enable furniture and nature to co-operate together into ultimate harmony. One uses the other, since they need each other to create the whole and final design. As a consequence, living qualities of nature will have a constant influence on their projects. They try to use human action as less as possible and to apply nature the way it is given. After creating a certain design, one can see that a part of nature is transformed into a sculpture, an utensil, without losing respect for nature.

Vienetta chair

Design Agency: Studio Floris Wubben

This chair is a combination of white polypropene with steel chromed pipe. By using these two materials an interesting combined action of contrast and construction arises.The construction of the chair can be taken apart; every part of the chair can be replaced. The polypropene is available in many colors.

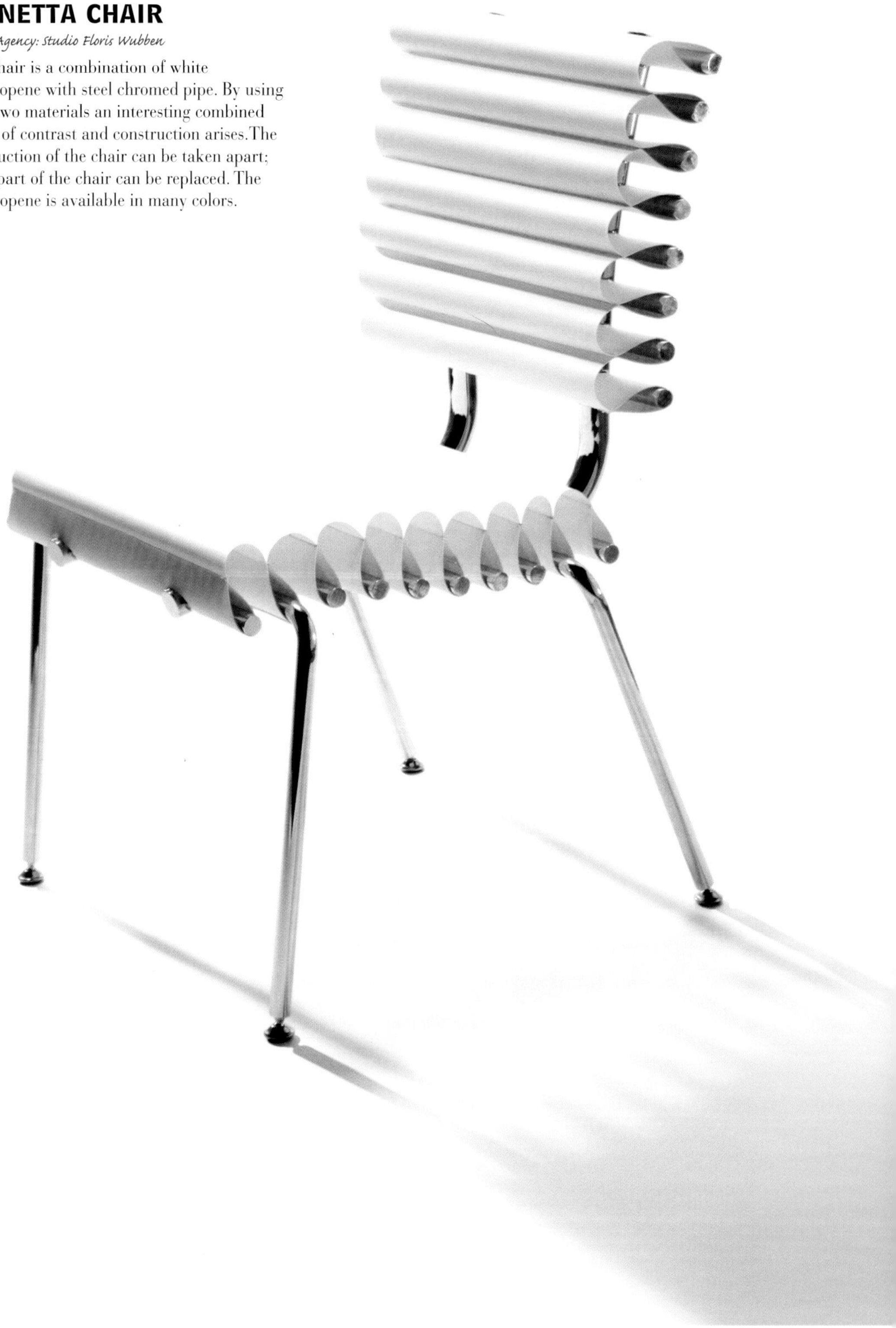

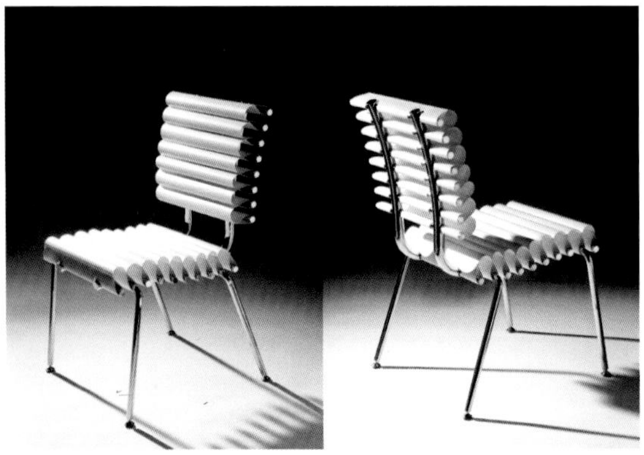

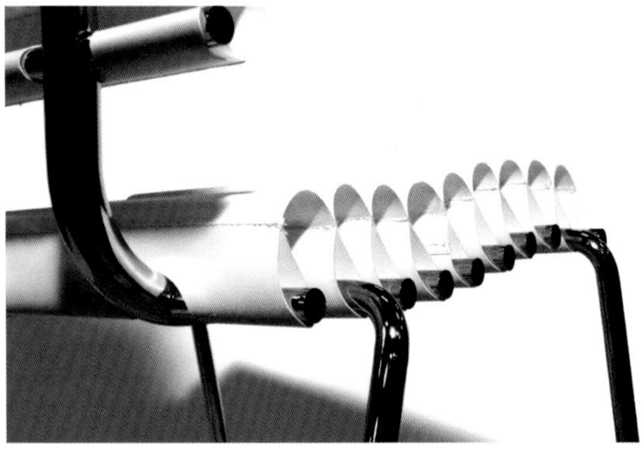

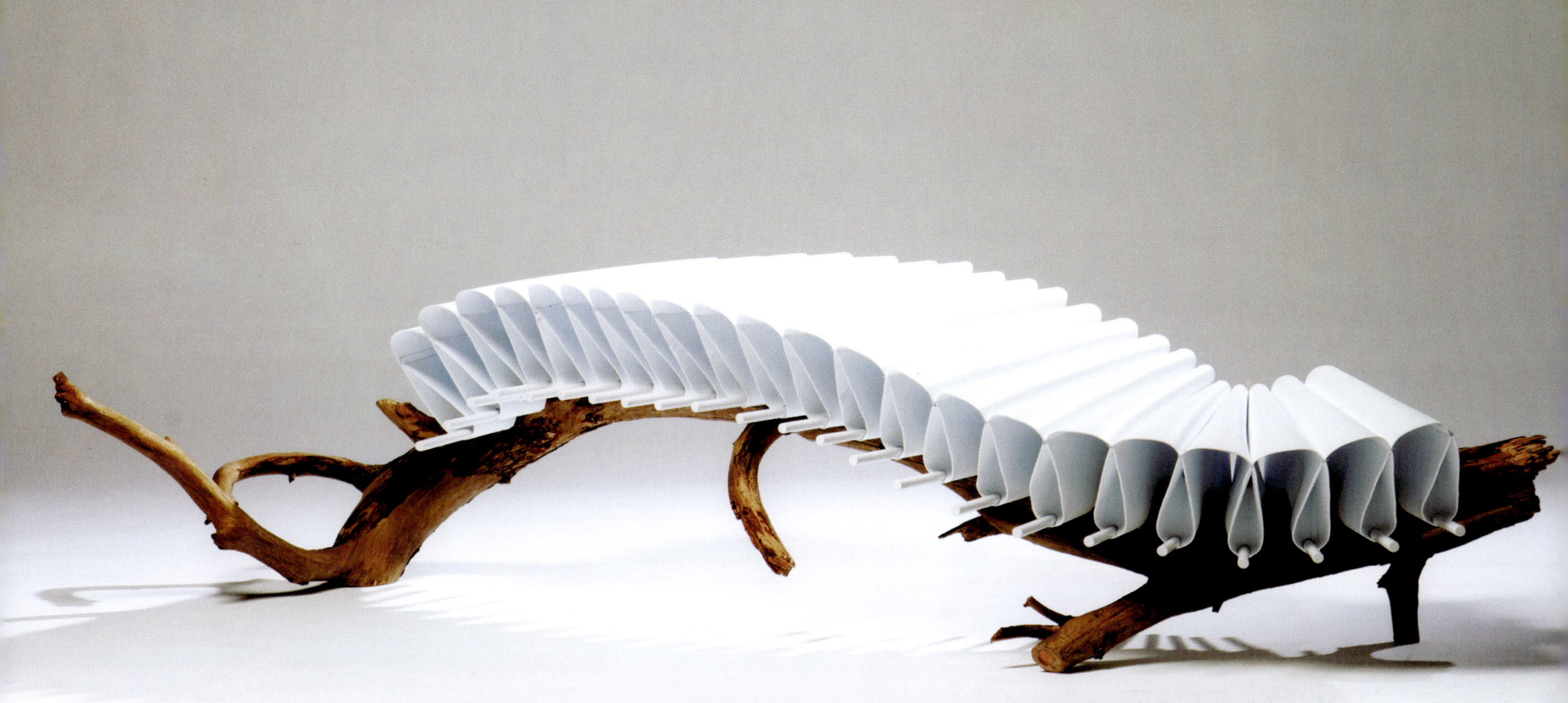

No. 3 BENCH

Design Agency: Studio Floris Wubben

This bench is made of polypropylene, wood and lacquered metal. The wavy polypropylene is attached to the wooden branch with metal rods. As a consequence, the shape of the polypropylene is given by the shape of the branch.

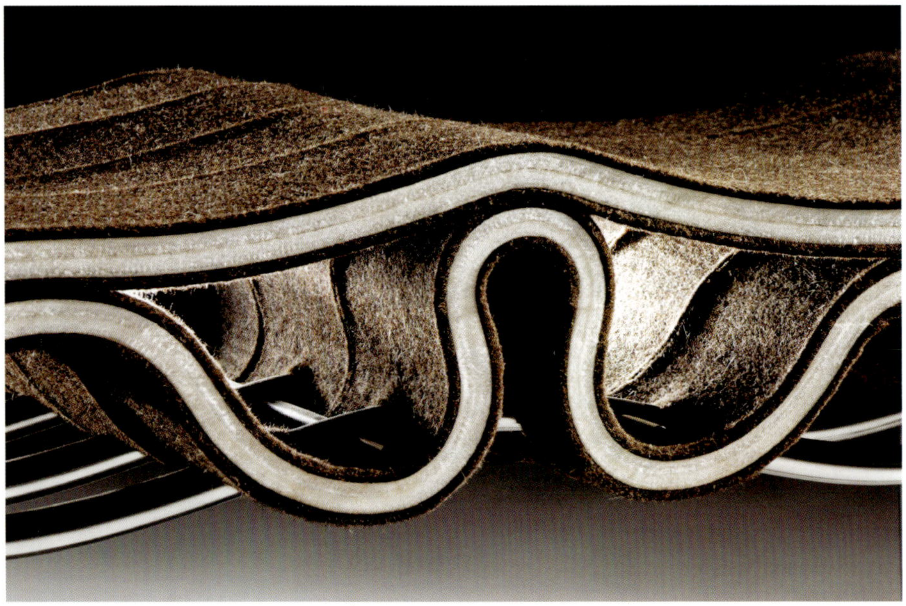

Plyfelt furniture

Design Agency: Studio Floris Wubben

This 'two seat bench' is a combination of felt (with 20 kg wood glue) with steel pipe. Through the use of these two materials an interesting combined action of contrast and construction arises. The felt can have many different kind of color combinations.

WertelOberfell Platform

WertelOberfell Platform was founded in 2007 by Gernot Oberfell (1975) and Jan Wertel (1976). Both studied Industrial Design in Stuttgart at the State Academy of Arts, a school which is based on the principles of the Ulmer Schule and the Bauhaus. Before forming their London & Munich based practice they both worked several years for Ross Lovegrove Studio.

Their work ranges from furniture, lighting and industrial products to experimental research pieces for companies such as Sony, Yamagiwa, MGX, Iker, BNOS, Koleksiyon, DuPont™ Corian®, Ferro, Profap and others.

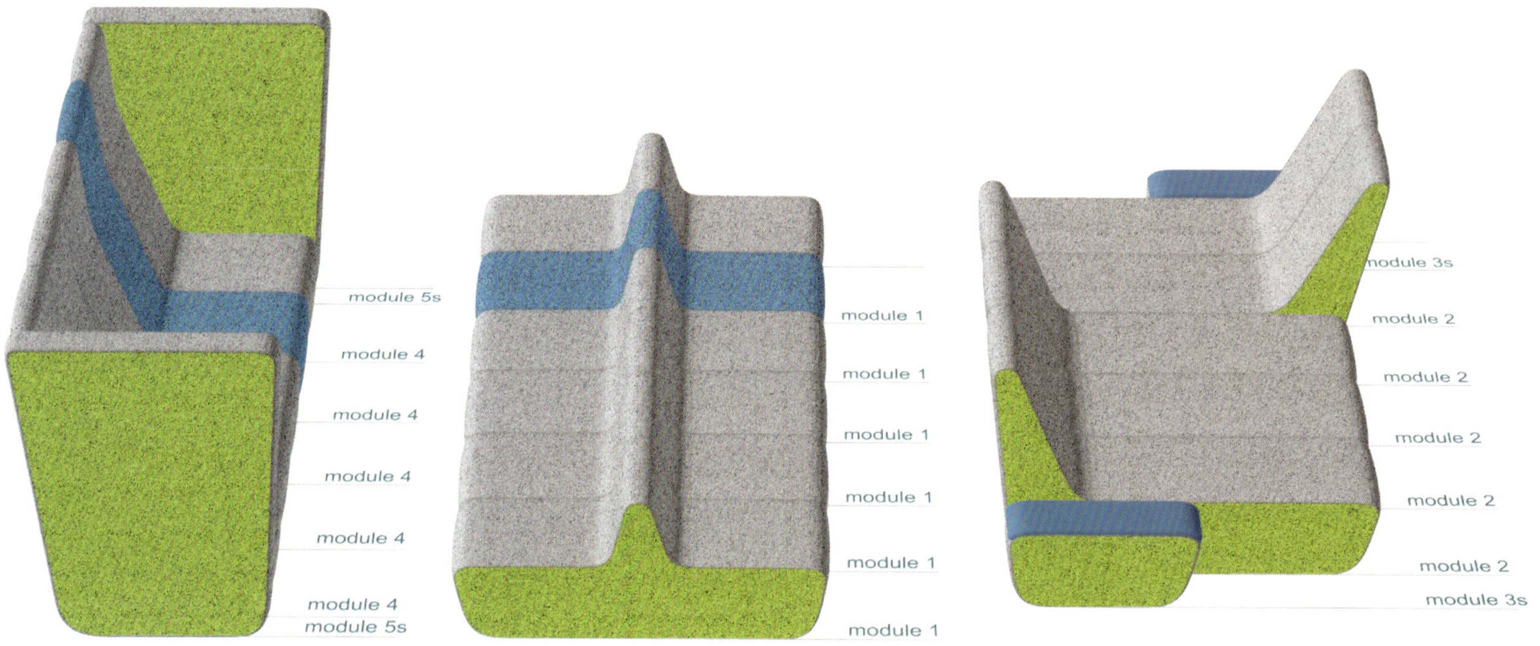
module 5s
module 4
module 4
module 4
module 4
module 4
module 5s
module 1
module 1
module 1
module 1
module 1
module 1
module 3s
module 2
module 2
module 2
module 2
module 2
module 3s

READING
READING
READING
READING

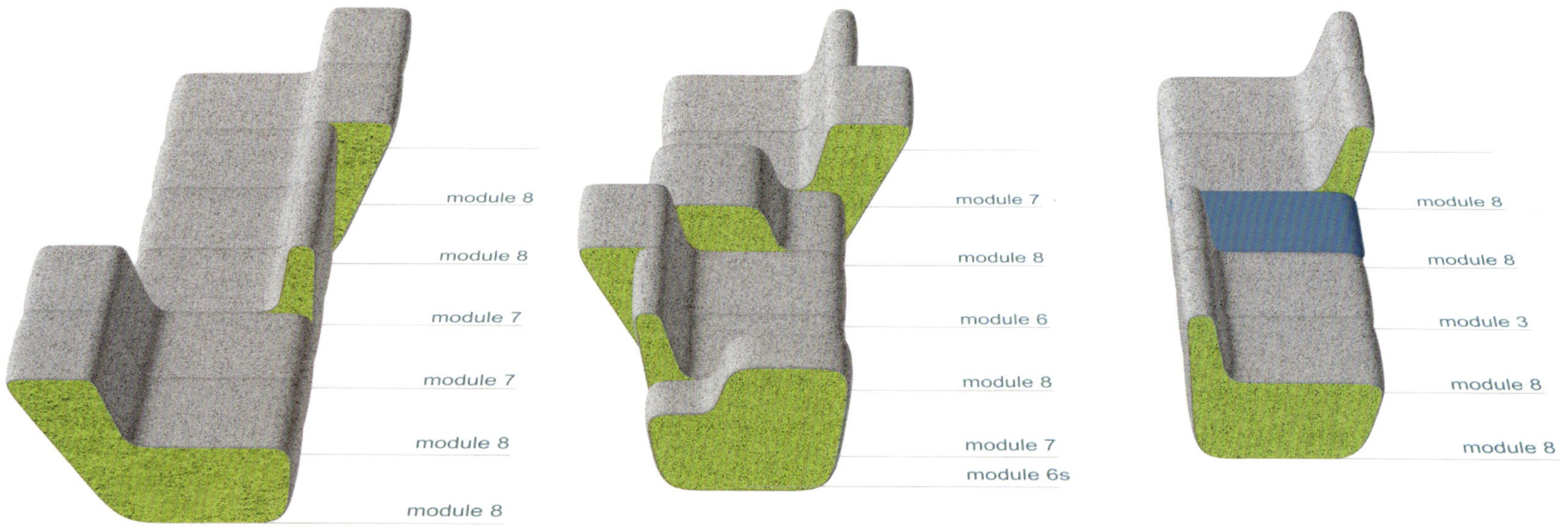

DILIM

Design Agency: WerteloberfeII Platform

DILIM is a modular soft seating system for the office and public spaces. The system can range from a very quiet and functional seating unit to a more exciting and expressive setup. DILIM consists of 8 distinct modules with different functionality. They can be arranged in many ways which make the system very versatile and adaptable. The colors of the horizontal textile stripes can be chosen from a broad palette and will help to tune the furniture pieces to work with the interior.

DZ Studio

DZ Studio was established in 2004 by Enrico Zanolla and Andrea Di Filippo. Since university, they showed a great attention for details, style, materials, animated by the desire of creating something new. Passion for design, creativity and technology are the features that best characterize their way of working, from the conception stage to the industrialization of the product. Their works were shown in the most important design exhibitions, cultivating originality in each project he adopts. Their objective is to tackle projects not only in terms of the aesthetic aspects traditionally associated with design, but by developing the entire creative process, including project conception, technical development, engineering and design.

Basil's triptych

Design Agency: DZ Studio

This project is a tribute to the beauty and uniqueness of the St. Basil cathedral in Moscow. During his stay in Russia Enrico Zanolla spent time observing and studying the constructive system of the domes. The difference in color and size comes from symbolic meanings assigned to every single dome. A great attention has been paid to details and quality of realized domes, and everything was made without using a 3D CAD.

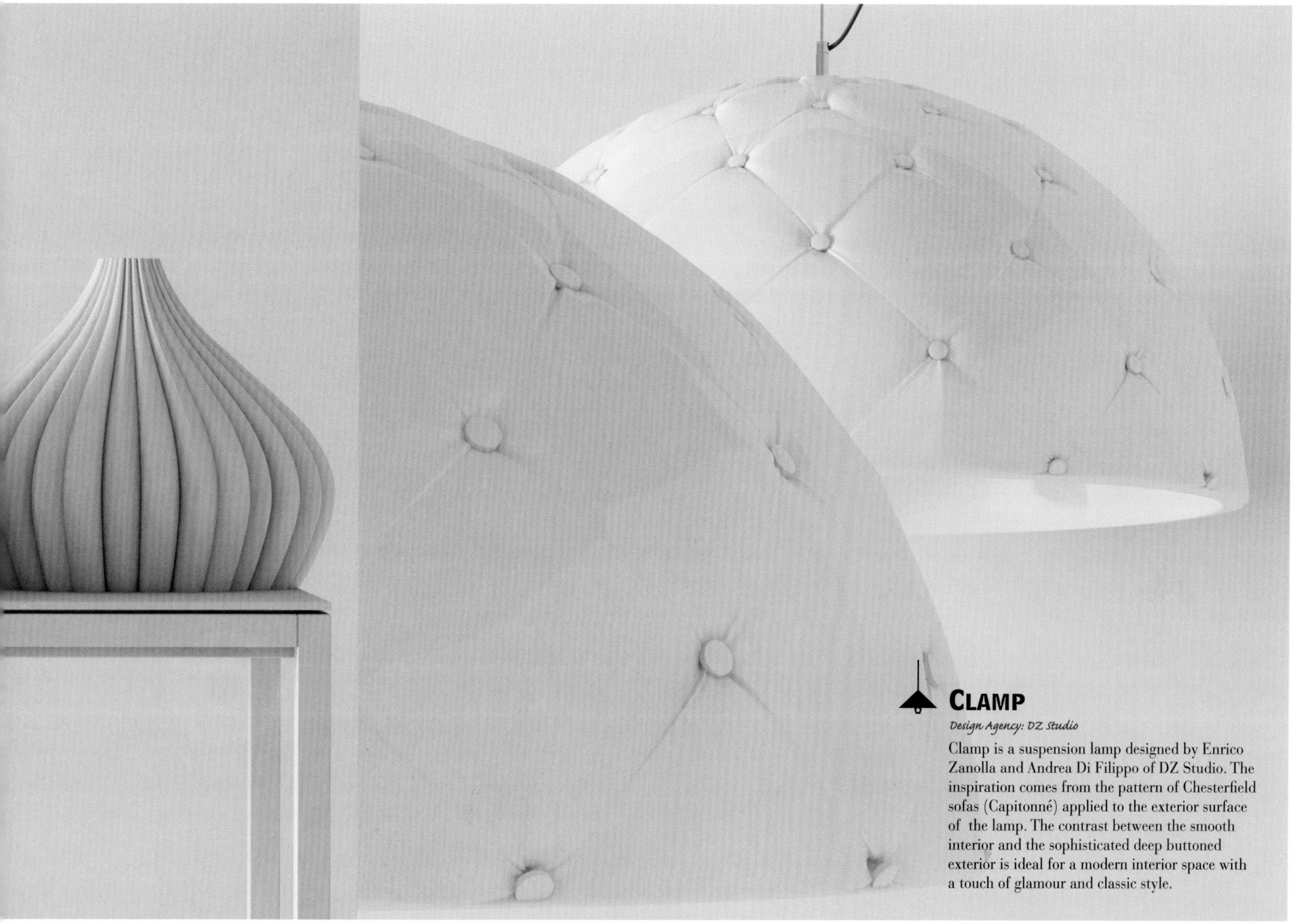

CLAMP

Design Agency: DZ Studio

Clamp is a suspension lamp designed by Enrico Zanolla and Andrea Di Filippo of DZ Studio. The inspiration comes from the pattern of Chesterfield sofas (Capitonné) applied to the exterior surface of the lamp. The contrast between the smooth interior and the sophisticated deep buttoned exterior is ideal for a modern interior space with a touch of glamour and classic style.

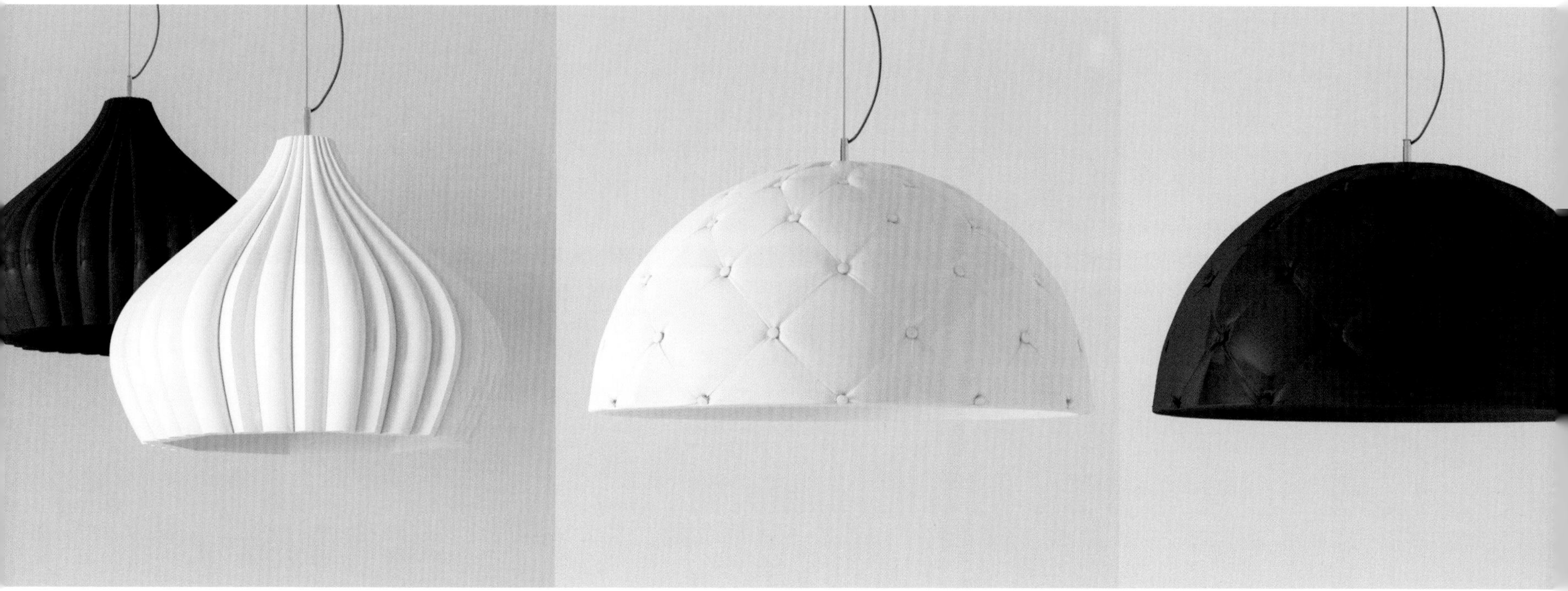

Sirio

Design Agency: DZ Studio
Dimensions: 100cm x 42cm, 45cm x 49cm

Sirio is a new collection of coffee tables and stool designed by Enrico Zanolla and Andrea Di Filippo for Miniforms Spa. The collection has been presented during the Milan Fair 2010 and it will be in production at the beginning of this summer. The painted metal structure is restrained by a circle giving the impression of being chocked. The coffee tables are available in two different sizes with glass top, one of which can have an upholstered seat becoming a stool. The structure is 12mm tubular welded steel, the top is 8mm glass.

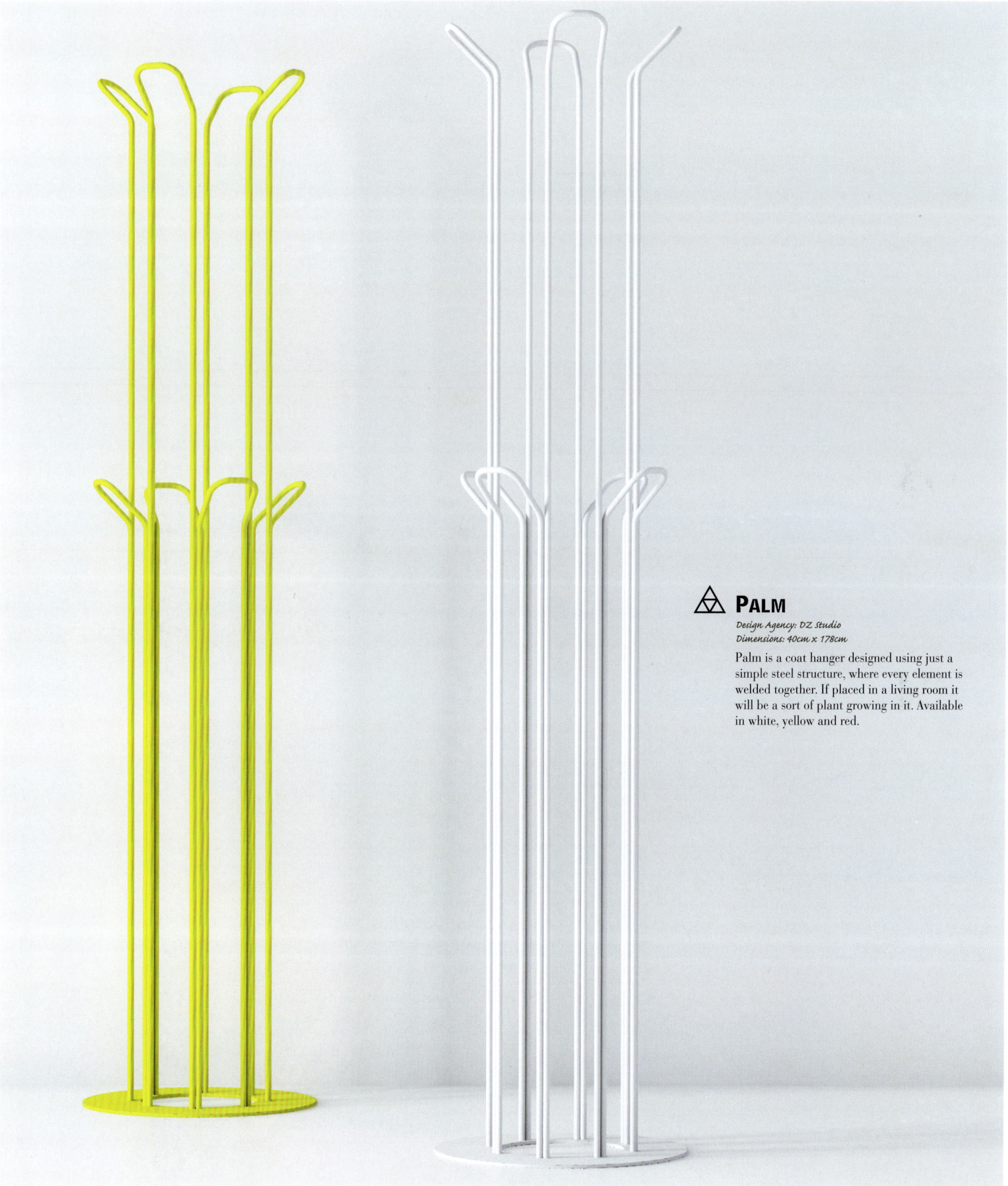

PALM

Design Agency: DZ Studio
Dimensions: 40cm x 178cm

Palm is a coat hanger designed using just a simple steel structure, where every element is welded together. If placed in a living room it will be a sort of plant growing in it. Available in white, yellow and red.

Joseph Walsh Studio

Over the last year, Joseph Walsh Studio has completed a number of significant commissions, including pieces for the Mint Museum, Duke of Devonshire Collection at Chatsworth house and most recently, the completion of the Enignum Canopy Bed as well as new work for upcoming exhibitions. Joseph Walsh Studio is looking forward to an exciting Autumn/Winter 2010 exhibition schedule in the USA and Europe.

Erosion Low Table

Design Agency: Joseph Walsh Studio

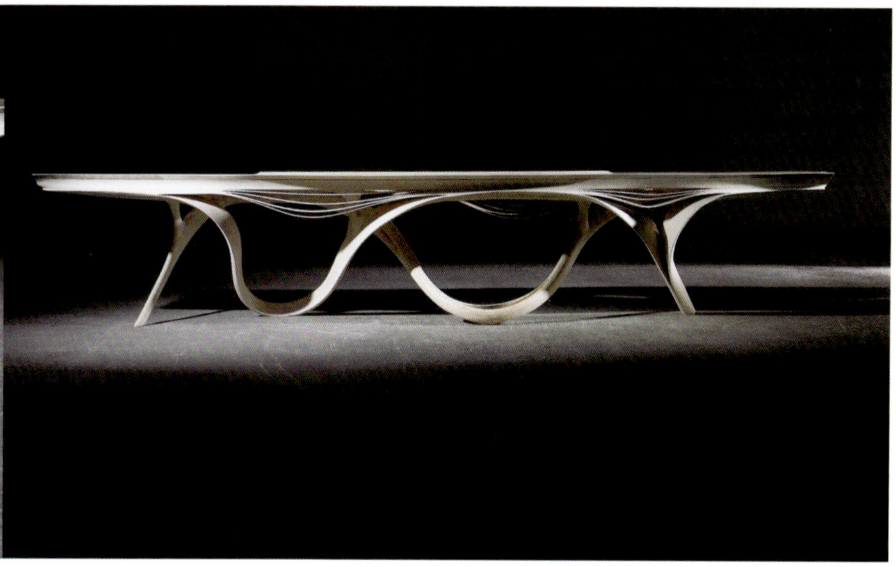

ENIGNUM TABLE

Design Agency: Joseph Walsh Studio
Description: Dining Table
Materials: burr olive ash, white ash

ENIGNUM CANOPY BED

Design Agency: Joseph Walsh Studio
Designer: Joseph Walsh
Dimensions: 250cm x 170cm x 194cm
Materials: olive ash and oak

The Enignum Canopy Bed is the first bed in the series. The concept behind the piece was to create a space, a cocooned feeling of protection and enclosure. The silk canopy creates a sense of enclosure which can be customised for each season.

ENIGNUM CHAIR I

Design Agency: Joseph Walsh Studio
Materials: ash and suede

AF Designs

AF Designs is a London based bespoke furniture company created by Italian designer & maker Andrea Felice. Work undertaken varies widely from sculptural freestanding pieces of furniture, to fitted furniture such as wardrobes and kitchens. AF Designs offers an exciting mix of traditional joinery techniques with contemporary design. Working with a wide range of materials from solid wood and veneer through to spray lacquered MDF, glass and metal. The will to experiment with materials—together with detailed cabinetmaking and carving know-how allows them to create unique and timeless pieces.

The Elementi Copper Fronted Cabinet

Design Agency: AF Designs
Photography: Decoratum

The Elementi Copper Fronted Cabinet was handcrafted in 2009. The front features patterned copper panels treated to give each panel a different texture and luster. Disguising 2 cupboards and 5 drawers. The unit body is constructed using hardwearing moisture resistant fiberboard & birch ply. Limited edition, made to order.

Pianca SpA

Pianca SpA is a well-established industrial company active in the field of furniture systems and complements for the home, for both the day-time and night-time areas.

The company was founded just after the end of WWII by Enrico Pianca in Gaiarine, in the province of Treviso, right in the heart of what would become the Furniture District. Gradually, from the 1950's to the 1960's, Enrico Pianca changed his business from an artisan workshop to an industrial organization. The company, which in 1965 already employed 160 staff, grew from a local concern to see development on an international level.

2010 DIVANI ATELIER VOL2

Design Agency: Pianca SpA

CAFÉ
ÉCOLE PRIVÉE
de GARÇONS

COIFFEUR

Spazio box

Design Agency: Pianca SpA
Dimensions: 360cm x170cm x 35cm/ 41cm/ 57cm

Floor-standing base units in Bianco Matt lacquered with Beta handle and plinth with aluminium finish. Unit: completed with flap door in Bianco Matt lacquered with front in Stopsol glass and frame with Brill finish. Wall units: completed in Terra Gloss lacquered with pressure opening system.

Fabrica

Fabrica is the Benetton Group's communication research center. It was created in 1994 with the aim of combining culture with industry and offering young people from around the world an opportunity for creative growth and multicultural interchange. Fabrica's artists develop cultural and social projects in the areas of design, visual communication, photography, interaction, video, music and publishing under the guidance of experts.

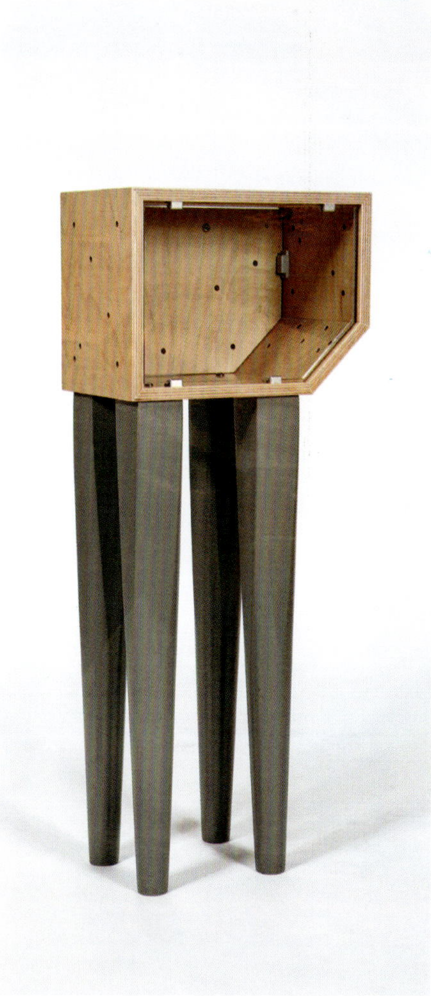

Down side up

Design Agency: Fabrica

Down Side Up is an encounter between experimentation and consumption, where design challenges shapes, volume perception and directions to create objects with constantly varying functions and interpretations. And so by adding a few essential accessories like legs, handles, hooks or hands, and by choosing a precise orientation, a simple basic wood unit is transformed into something new with a unique personality. The heart of each unit is a hole, a functional element which also gives the unit an aesthetic value. Therefore the effectiveness of each article lies in its versatility of use and its potential for creating customised modular systems.

KAMKAM

Kamkam means 'extreme darkness' in Korean. The darkness that is made of millions of colors represents the design of KAMKAM's members: Hyunjin Seo and Jaekyoung Kim. KAMKAM makes space that people can freely conceptualize, construct and share anything with others. KAMKAM transforms all this darkness into every single one of its colors.

AN FURNITURE

Design Agency: KAMKAM

Ax sheets (A3, A4, A5, etc) have 'Silver Ratio' that promotes productivity by keeping the dimensions always (approximately) in the same ratio of width:length when they are folded in half. Inspired by these Ax sheets, we assigned fresh sculpture of distinctive ratio to furniture, which makes the furniture economical by saving raw materials. Especially, 'An Furniture' is convenient for single-person households since it also can be rearranged and relocated to a closet, a bookshelf, a storage cabinet and a drawer by a straightforward process.

Alexander Pelikan

'By designing I want to explore new roads and I want to be surprised where these roads take me.' Experimentation with common and exotic materials and techniques for example is such a new path. Another one is working with interesting craftspeople, other designers and innovative companies. While wandering these ways ideas emerge which are transformed and integrated in Alexander Pelikan's works. In Alexander's designs he is embracing coincidence – any detail could be of importance.

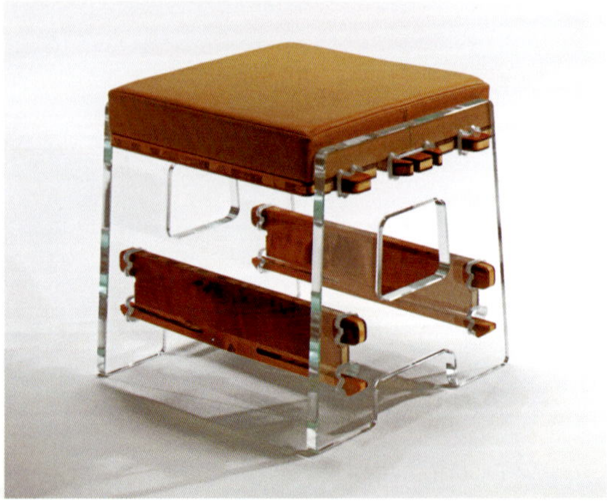

Clicfurniture Glass

Designer: Alexander Pelikan

What happens if you make the whole thing from only one material? What is needed to achieve this?
The answers lead to the Cliclounger, they were as follows: the construction has to consist only of the material, thus be made of this material. The material has to be finished right away. Alexander Pelikan constructed a chair out of Trespa, an HPL material very dense and strong with a durable top layer, developed for Ahrend and now mainly used for facades of buildings.

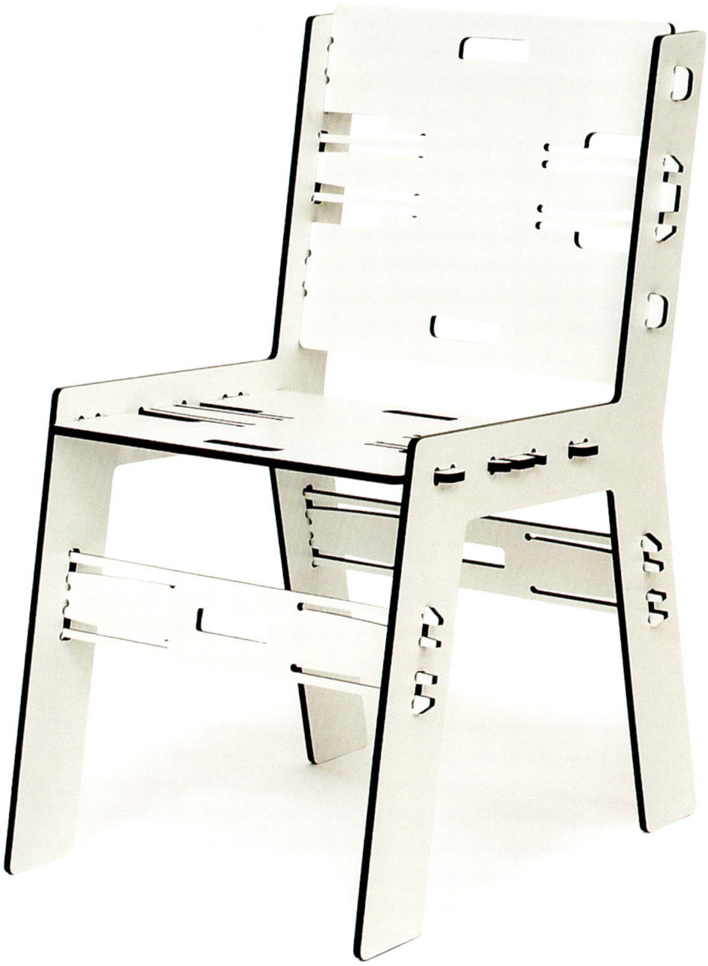

The clic diner chair

Designer: Alexander Pelikan

Materially, the clic diner chair defines current notions of sustainability. It is composed of bamboo harvested from a sustainable bamboo forest, where only the ripe shoots are collected and the rest of the forest is left to mature naturally. The chair is produced in a caramel-colored, heat-treated bamboo plate material. Heat is applied to the material in order to caramelize the natural sugars within the bamboo, thus making it less appetizing for insects and mold. The final product requires no additional finishing, though wood oil can be applied for a more polished look.

THE PLASTIC NATURE CHAIR

Designer: Alexander Pelikan

The Plastic Nature is a furniture concept which exemplifies the connection of the 'plastic world' and the world of wooden furniture.

The curiosity kitchen

Designer: Alexander Pelikan

Curiosity Kitchen is a space, more than just a piece of furniture it gathers functionality of its worktop with manifold drawers and cavities from which highly refined ingredients, special liquors and rare spices come out as from a magic hat.

Fishbol Design Atelier

This workshop-based studio was formerly established in Ottawa (2005) by design duo of Elie and Mani—both graduates of Carleton School of Architecture. Although strong ties have been kept between Ottawa and Toronto, collaborations have stretched to Montreal, Seoul, London and Kuwait. This workshop specializes in extremely original, high-impact projects for architecture, interior, event installations, furniture, product and fashion design.

Wine-bottle shelf

Design Agency: Fishbol Design Atelier
Designer: Mani Mani
Dimensions: 40cm x 120cm x 2.5cm
Material: kirei board

Kirei board from reclaimed stalks of the sorghum plant or finished in white lacquer, and recycled wine bottles.

Bungeeseat

Design Agency: Fishbol Design Atelier
Designer: Mani Mani
Dimensions: 60cm x 70cm x 60cm
Materials: bamboo plywood, assembled bungee cords with hooks.

The Bungeeseat revisits René Herbst's Chaise Sandows, but adds a contemporary twist by using today's aesthetics and conscience in material selection. The frame is constructed from bamboo plywood (a rapidly renewable alternative to wood) and locally manufactured bungee cords. The use of bamboo plywood with triple layered perpendicular bamboo parts has added extra strength to the structure of the frame and allows for more design flexibility compared to conventional plywood. The Bungeeseat kit fabricated in Fishbol's Toronto studio, comes in customizable accent colors and bungee cords. The ready-to-assemble (RTA) chair is sold in a neatly designed package which disassembles to become re-assembled into the chair.

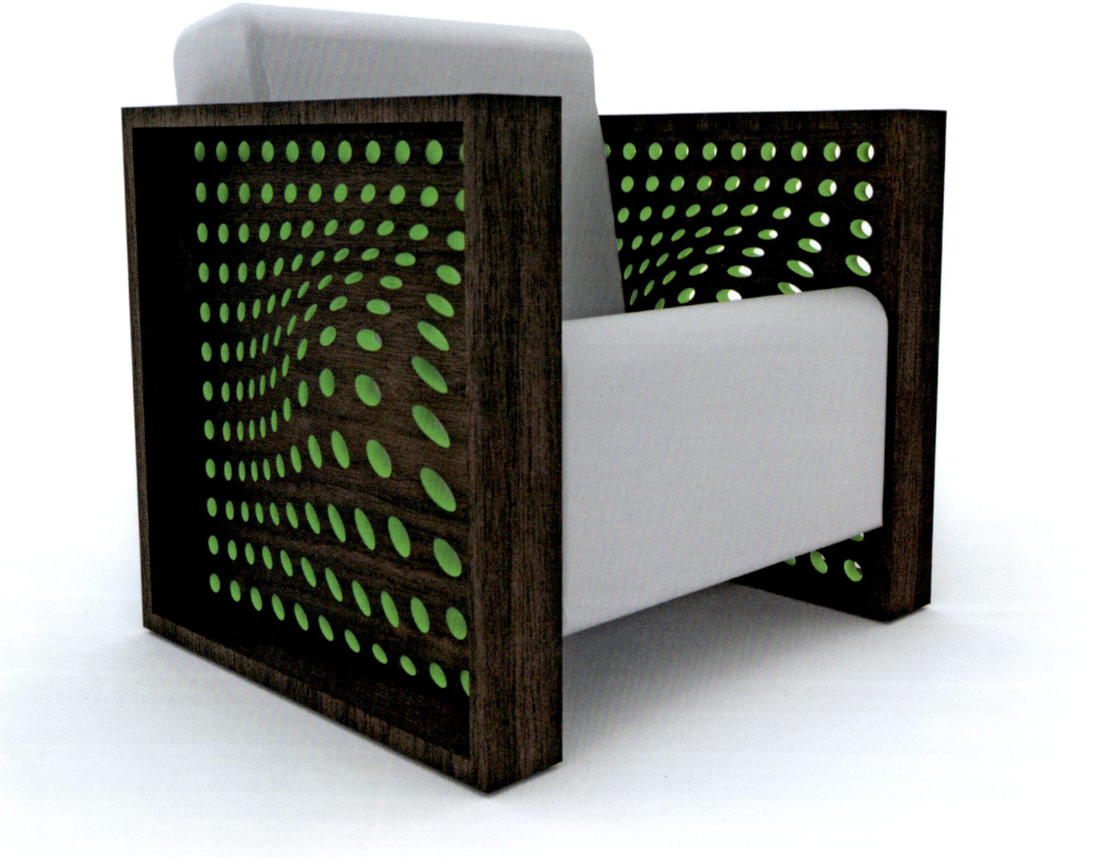

Gazebo chair

Design Agency: Fishbol Design Atelier
Designer: Elie Nehme
Dimensions: 650mm x 650mm x 800mm
Materials: Walnut + high-density foam

'Gazebo', it incorporates a soft linen seat with two oversized panels fabricated out of dark walnut, framing a pattern of chartreuse-colored holes. The study of patterns begins with a uniform grid of circles that is digitally manipulated, morphing the size and shapes of the circles, so that a 2D surface begins to attain a 3D quality. The resulting visual effect is the weight of a body sitting and the reacting forces left as impressions on the oversized panels, thus giving the chair a dynamic living quality to it. The Gazebo combines the nostalgic idea of a leisure chair and material, with an exploration of spatial effects that can be achieved with contemporary techniques.

Lobstertrap chair

Design Agency: Fishbol Design Atelier
Designer: Elie Nehme
Dimensions: 900mm x 900mm x 850mm
Material: veneered plywood

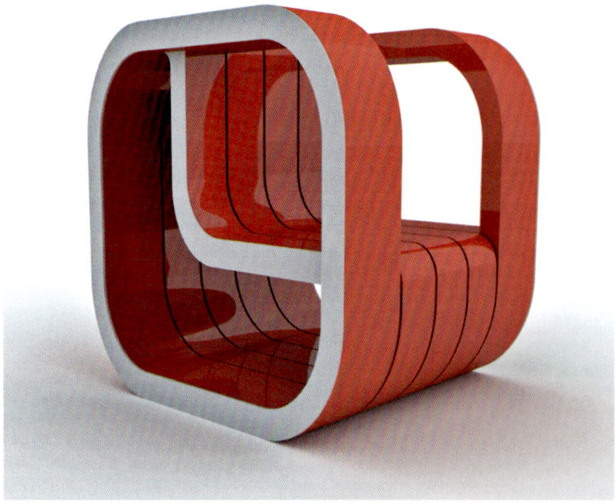

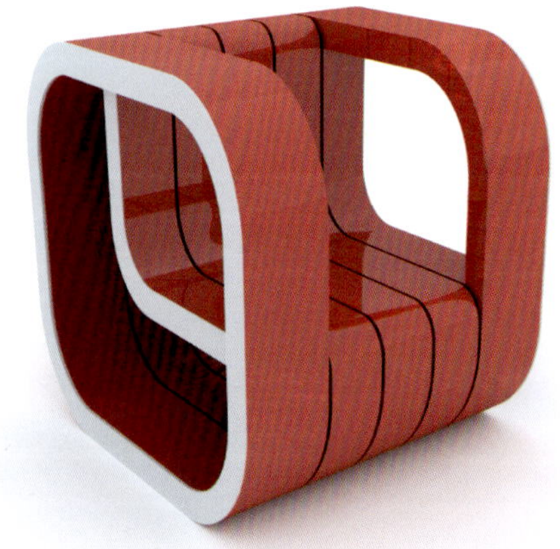

The Bookseat

Design Agency: Fishbol Design Atelier
Designer: Mani Mani
Dimensions: 600mm x 920mm x 843mm
Material: plywood

The Bookseat is a simple bookcase that playfully curves and becomes a seat; is a creative design responding to the advent of multifunctional spaces in today's urban living. Noted as one of the 'ten must sees' at Toronto Interior Design Show 2008. The Bookseat is hand-crafted in Ontario and available with a felt cushion in customizable colors and limited edition leather cushion.

Jessica Carnevale

Since graduating from the Rhode Island School of Design in 2004, Jessica Carnevale has designed furniture and products for clients around the world. For the past two years, Jessica has been working at the Marcel Wanders Studio in Amsterdam. Under his direction, she has developed collections for Baccarat, Target, Bardelli, Philips, Xo, Magis, Cappellini, Skitsch and Moooi.

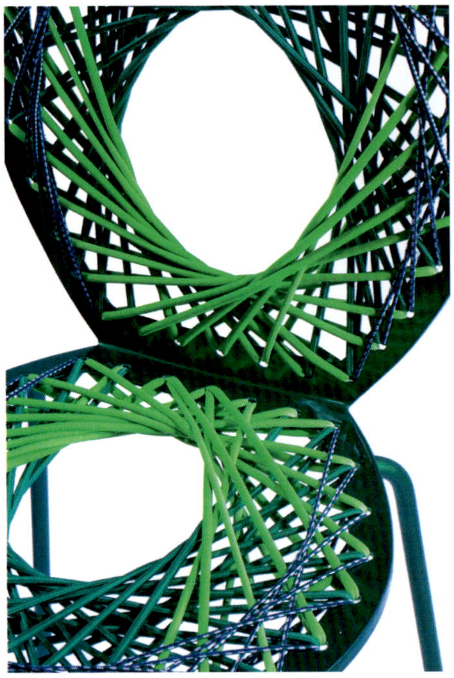

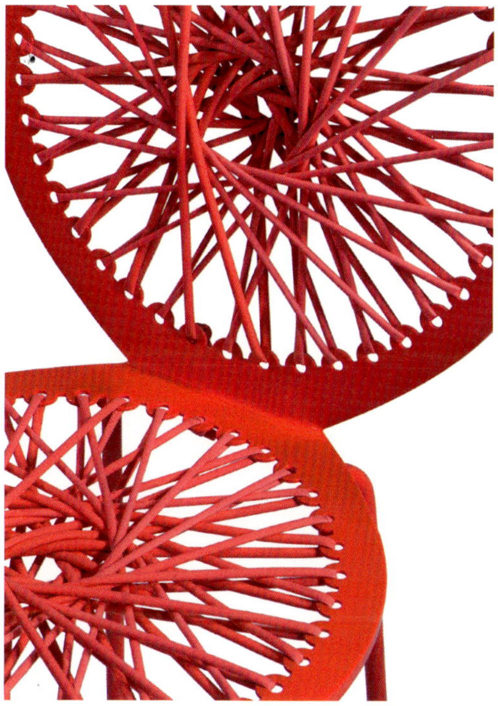

Stretch chair

Designer: Jessica Carnevale

With her debut collection 'Stretch', Jessica wanted to bring some of the exuberance of the fashion world into her designs. She experimented with a number of different materials and chose latex, bungee and rope for their bright colors and wondrous elasticity. She built a simple chair to better frame the compositions that she created. The result is a collection of three vivid pieces, after all, design should be beautiful and useful, but also fun.

Rui Alves

Born in Paços de Ferreira, Portugal, in 1977, he comes from a family of skilled craftsmen; his grandfather was a carpenter and his father a woodworker. Since his childhood he has lived surrounded by tools, materials and his mind always working on creating or disassembling something. Rui studied Industrial Design at IADE (Instituto de Artes Visuais, Design e Marketing) in Lisbon graduating in 2001. In the same year he set up his own studio (MY OWN SUPER STUDIO), back in his hometown, dealing with projects from furniture to product design. His approach to design reflects his love for materials, processes and for simple everyday objects. Always paying attention to details, almost an obsession, he tries to make it simple, practical and playful and at same time give it a personality.

Fiss Family

Designer: Rui Alves

It's a group of four lamps that can be located anywhere, either individually or, preferably, as a family. Each lamp is made of wood and has a smooth lacquered interior. Each has a very pleasant light quality that can be enjoyed as a singular or a grouped experience.

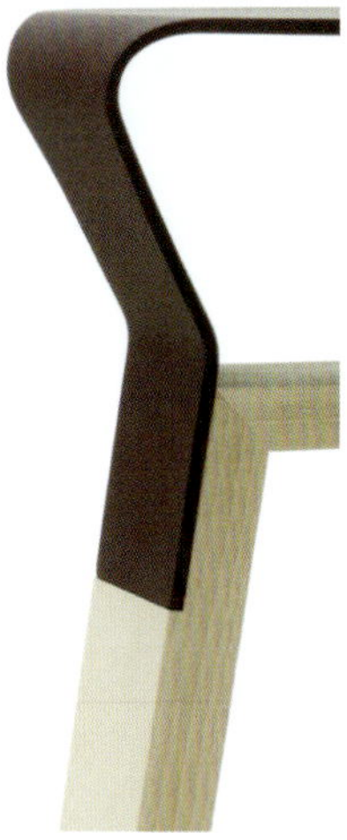

Avo

Designer: Rui Alves

It's a small tribute to Rui Alves's grandfather. He used to have his favourite wooden stool and Rui remembers it so well, because he would keep it handy for almost any occasion. Made from anodized aluminium and solid wood it's yours and perfect to use everyday just as you please.

Miu

Designer: Rui Alves

With just a few shelves and a number of the colored components it is possible to you play and create a structure that is light and easy to assemble. The support element is made of aluminium painted in a variety of colors and the shelves are made of wood.

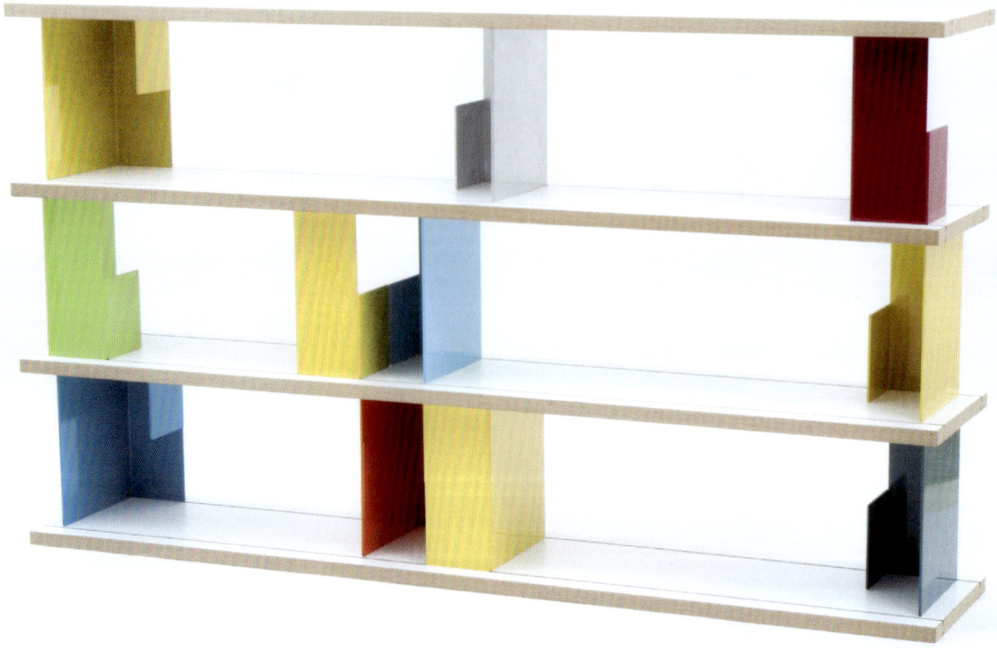

Welcome to the Jungle

Designer: Rui Alves

James (elephant), John (hippopotamus), Jeremy (lion), Jerry (giraffe) and Joe (crocodile). Five friends that you can use almost anyway you want – you just have to catch them first. As tables, stools, or even coat hangers. They can even jump over one another in order to make some shelves. Take one home. Take them all home, go wild and enjoy!

Giancarlo Zema

Born in Rome in 1973 and graduated in Architecture at the 'La Sapienza' University of Rome, demonstrating since the beginning a strong futuristic style that attracts thanks to its intimate connection with Nature. New living solutions on water start to grow: being the result of a bright creativity combined with rigorous method, they internationally become an example of new ways of living. Thanks to a continuous and purely personal research—fed by a great passion for architecture and love for soft morphological shapes—architect Zema explores new technological horizons till he bravely melts architecture and design together, originating innovative projects that are outside the norms.

Leaf collection

Designer: Giancarlo Zema

A sinuous, flexible and soft leaf bending on itself: architect Giancarlo Zema was inspired by this vision to create his new extraordinary design collection of living solutions—table and chair—for Avanzini Group. Dupont™ Corian®, a contemporary material, shapes precious unique pieces, brightly colored, like the flexible-backed chair that fit itself to the body .

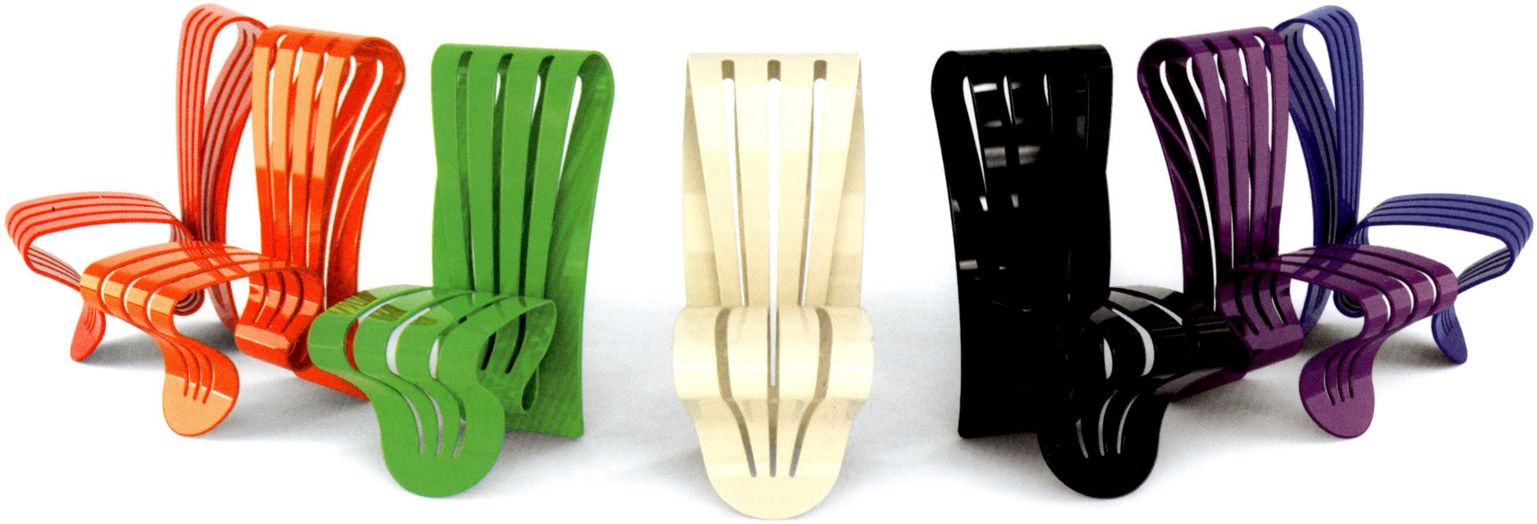

Valentina Gonzalez Wohlers

Product and interior designer based in London. Before entering Central Saint Martins to complete a MA in Contemporary Furniture (London 2005-2007) she graduated in Industrial Design with honours from the Ibero-American University in 2000, Mexico City. Her work has been featured in a wide range of international publications and media (Once Upon a Chair by Die Gestalten Verlag, 2009, Architectural Digest, Wallpaper*, Vogue Living, cover of Marie Claire Deco, among others) and selected for design exhibitions around Europe, where Gosserez Art Gallery in Paris and L'eclettico in Milan also represent her work. She is a regular guest lecturer and speaker in several design colleges throughout the UK. Currently a professional furniture and interior designer and consultant, Valentina has been developing private commissions and conceptual furniture for the residential and commercial industry in Europe and Mexico since 2007.

The Prickly Pair Chairs

Designer: Valentina Gonzalez Wohlers

The Prickly Pair Chairs combines the distinctive formal and aesthetical values behind Mexican and European figurative idiosyncrasy resulting in a furniture object within which both cultures are identified. It is at once frivolous and humorous yet in the same moment encourages reflection and acceptance. Shapes, materials, colors, textures and finishes collide in irreverent and playful harmony: Prickly Pair Chairs is a statement against prejudgements and preconceptions.

Måns Salomonsen

Måns is a 24 years old Swedish design student. He is a hands-on person full of ideas. With his patient and hard working personality he's the person who works with every project until the very end. Måns's mature, open and curious attitude for design gives his works a specific character and combining that to his handcraft skill and knowledge, nothing is impossible. He has studied a two years foundation course in cabinetmakingand at 3 years BFA education in Woodoriented Furniture Design.

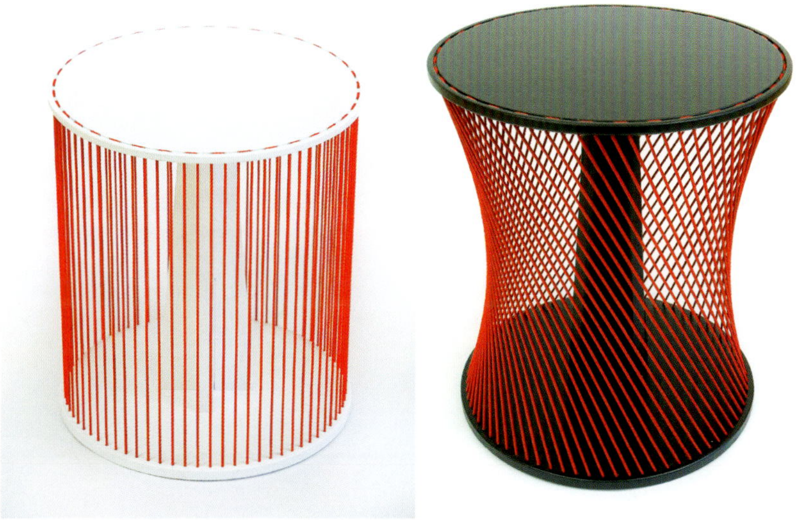

Playful

Designer: Måns Salomonsen

There is a mechanism under the seat with two axis that create the movement of the seat. No screws or glue were used in its construction. It is all held together by the elastic band and the weight of the person using it. The band makes the seat surface bounce back to a horizontal position when it is not in use. The elastic also create a see-through surface that gives the stool volume and the option to create different looks.

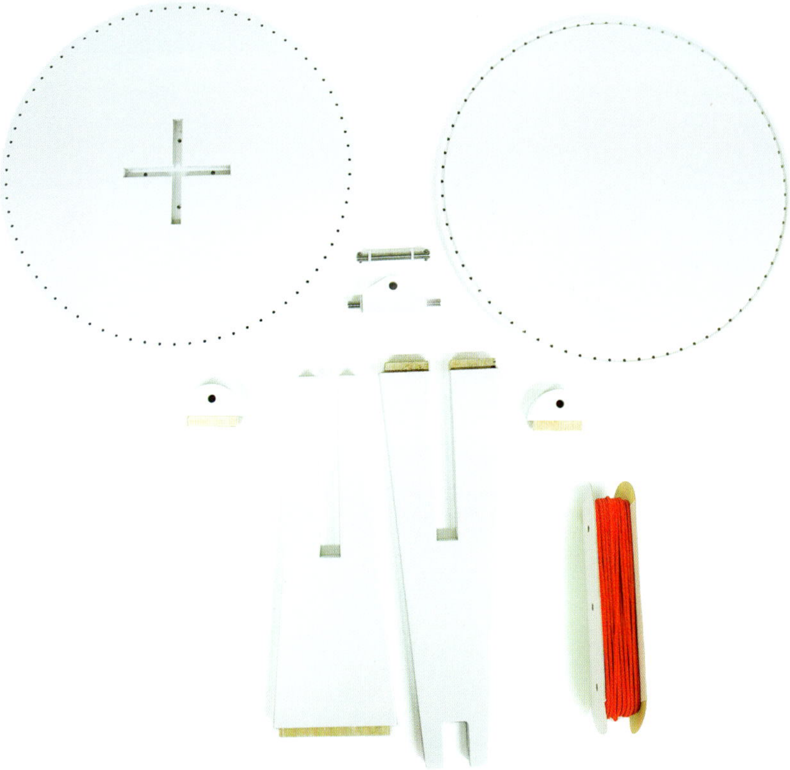

Bar table

Designer: Måns Salomonsen

The construction is made of solid metal bars in the under frame that both supports the tabletop and works as a foot stand and the tabletop is made of aluminium foam, filled with black-colored araldite plastic to create a patterned surface. Around the aluminum filled the surface is a stainless steel edge that protects it from bumps and scrapes.

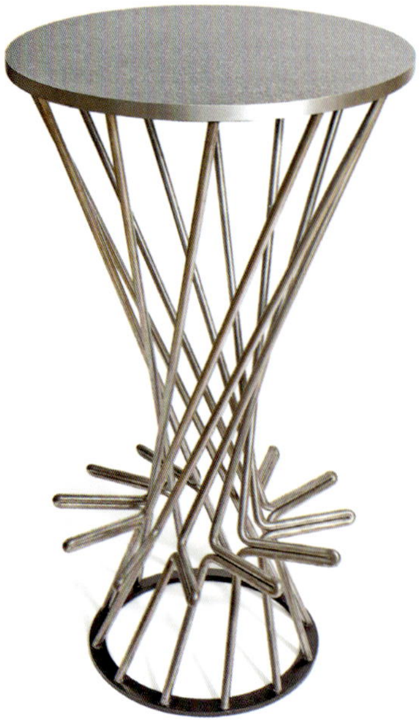

SPINE

Designer: Måns Salomonsen

This piece is the result of a project where Måns Salomonsen had to design and make storage furniture with at least two drawers. Måns chooses to challenge himself by trying to combine wood and metal into a modern and elegant design. One part of his idea was to try and play with the illusion of balance.

LUND & PAARMANN

The Danish design studio LUND & PAARMANN was founded in 2006 by the two designers/ architects Oluf Lund and Eva Paarmann. Oluf has a background as a carpenter before he graduated from The School of Architecture, Aarhus, Denmark. Eva has her diploma from the same school, and since then she has specialized in 3D-modelling. The combination of IT and practical skills gives the couple a wide range of possibilities when it comes to making the most of their creativity. The ambition of LUND & PAARMANN is to make new designs which make sense regarding the environment, function, construction and aesthetics. Rule number one: there is no need to repeat what has already been done.

THE LOBSTER CHAIR

Design Agency: LUND & PAARMANN
Manufacturer: Strictly Design A/S
Materials: laminated wood, leather and steel.

The Lobster Chair was introduced at IMM 2008 and since that it has obtained international acknowledgement as 'a new classic'.

The Zquare Lounge Chair

Design Agency: LUND & PAARMANN
Manufacturer: LOPFURNITURE
Materials: laminated wood, steel and solid wood.

The Zquare Lounge Chair is a classic Scandinavian design. The main idea and construction is similar to the Zeeen range, but the appearance is quite different.

The Zeeen Range

Design Agency: LUND & PAARMANN
Manufacturer: LOPFURNITURE
Materials: laminated wood and steel.

Zeeen is a flexible furniture concept, which gives the user the possibility to 'dress' the chair according to her own needs and taste. Due to the construction principle the seats are very strong and the designers have obtained double-curved seats by using single-curved components.

Troll Family

Design Agency: LUND & PAARMANN
Manufacturer: LOPFURNITURE

The Troll Family is a small range of furniture, inspired by Scandinavian troll tales. The materials are solid oak, full grain saddle leather and Icelandic sheepskin.

Werner Neumann

Werner Neumann loves the lightness of form and life of materials. He admires people who dare to embark on new paths and can touch you through their enthusiasm and courage. In Werner's own collection he also looks forward to the challenge of the unknown, new materials, different shapes, but always in combination with functionality.

BIRCHWOOD COLLECTION

Designer: Werner Neumann

With birch bark that he took in the woods, the designer Werner Neumann from Eindhoven, made a series of furniture covered with bark; the Birchwood collection, including a dining table, lamp, chest of drawers and his latest design: the birchwood armchair.To get the bark surface, it is pressed and glued on wood. Some pieces are treated with epoxy and then painted, other models remain rough.

Robert Brou

Robert Brou is a furniture maker for hire with 15 years experience building everything from Chippendale to Nakashima. His original artistic creations explore the random beauty of nature through sculptural organic furniture.

Originally from New Orleans, he spent almost six years training in refinishing and building high end antique reproductions for two of the finest antique galleries on Royal St. in the French Quarter. Upon relocating to Atlanta in 2000 Brou opened his own one man shop, Naturalism Furniture, later that year. Over the past decade he has created hundreds of custom creations for clients as well as developing a unique style with his organic original designs.

'CATERPILLA' BENCH

Designer: Robert Brou
Dimensions: 90cm x 38cm x 45cm
Materials: poplar, pine, walnut

The fun 'Caterpilla' bench is large enough to fit two adults yet has a minimal footprint. The seat is made of Poplar lumber with Pine legs and Walnut tentacles. The reclaimed Pine comes from a burned building in Atlanta's West End and a factory in South Carolina. There are twenty-six legs, thirteen per side. The seat is contoured for comfort yet shallow enough to allow the piece to function as a coffee table as well.

'WAVE' DESK

Designer: Robert Brou
Dimensions: 198cm x 60cm x 78cm
Materials: poplar, MDF

Under the slab top the desks two towers are clad in 2 ½" thick Poplar lumber that is carved into a random 'wave' pattern that is created spontaneously by the artist during fabrication insuring each work will have its own unique personality. Between each layer there is a gap measuring three quarters of an inch tall, just large enough to fit adult-sized fingers, simply grab an individual section and pull the drawer open. 'The innovation of sculptural cabinetry that lies outside the 'box' is a liberating feeling.' – Robert Brou.

Boxetti

The creator of the Boxetti collection is Rolands Landsbergs (1976). He was born and rised in the artists' family and now resides and works in the Jurmala City, Latvia.
Since 1998, Rolands runs his own design and manufacturing company EL Studija (www.elstudija.lv), specialised in furniture, showroom and exhibition stands design and production for illustrious clients.
The Boxetti collection by Rolands Landsbergs was first presented internationally at The 100 Percent Design exhibition, London—the UK's leading contemporary interior and architecture event.

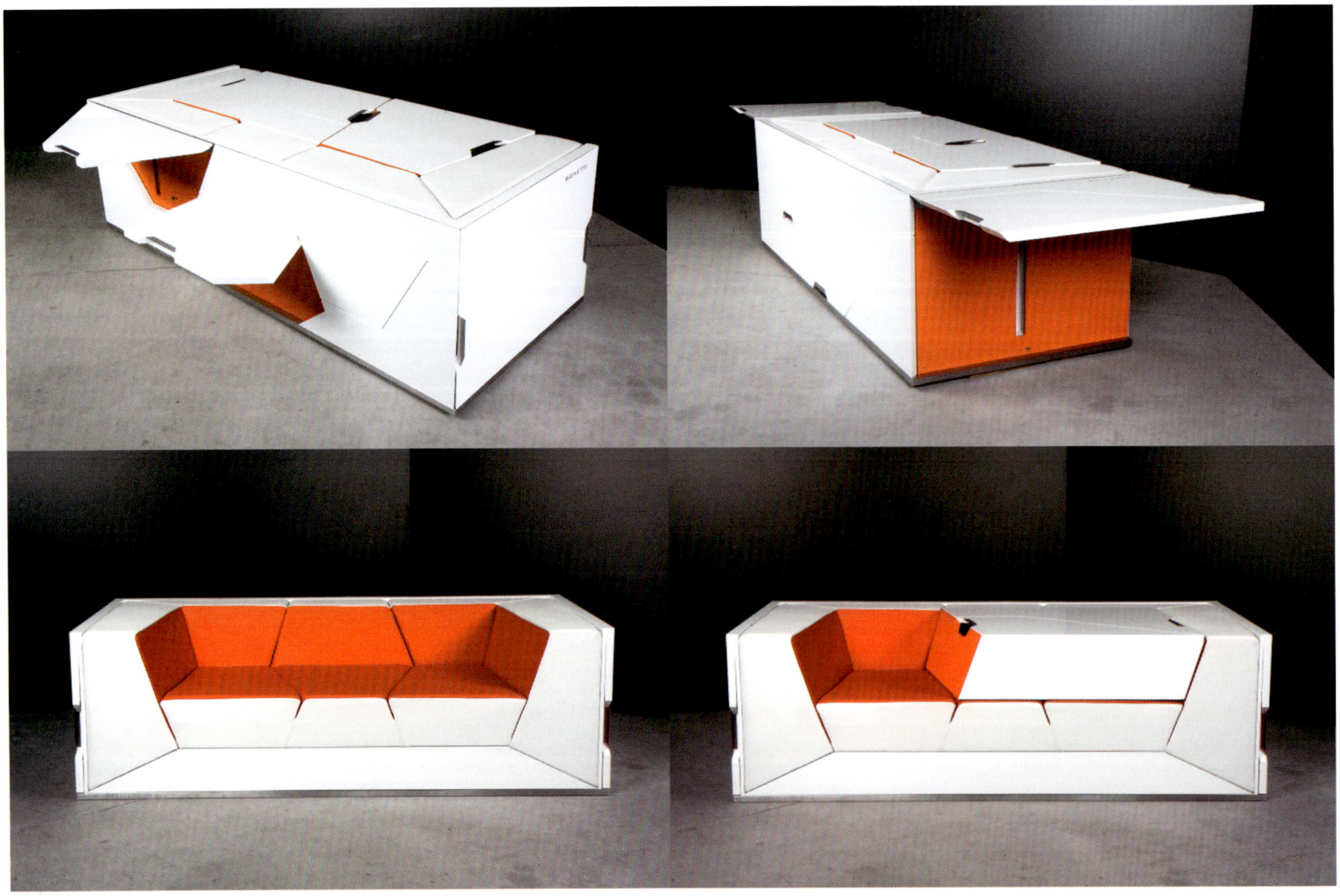

Boxetti Lounge

Design Agency: Boxetti

Boxetti Lounge contains the set of integrated furniture for a living space—a unit of three-seat couch with removable blocks of occasional tables. The compacted module can be used as a freestanding table surface. The unit has a foldable side panels to extend the top surface. The couch is upholstered and can be finished with leather or fabric up to demand. The unit is equipped with a small additional workspace consisting of foldable desk top and a seat surface behind the couch.

BOXETTI

Boxetti Lunch

Design Agency: Boxetti

Boxetti Lunch is a multifunctional kitchen island unit containing all the necessary kitchen equipment. The unit has an automatic sliding surface uncovering a built-in sink with a tap. The sliding surface is equipped with a safety sensor to stop closing automatically if any hand movement is detected.
The work surface contains a closed space for tableware together with a default set of kitchen knives. The part of the top surface is foldable to uncover the oak tree cutting board, the foldable cover has a built-in LED spotlight. The unit is equipped with two extensible bar seats, a space for built-in fridge, a standard set of shelves and drawers.

EXPORMIM

EXPORMIM is a company with great tradition in the production of furniture. With 50 years of experience, the company has consolidated its market thanks to the ongoing process of modernization, with special emphasis on the improved design of their manufacturing processes, the selection of raw materials and a careful design of their products.

EXPORMIM CHAIR SERIES

Design Agency: EXPORMIM

CHAISE LONGUE AIR

Design Agency: EXPORMIM

Nido

Design Agency: EXPORMIM

An armchair or bench seat, 'Nido' is built of 6mm steel tube and rods. Spanish designer and teacher Javier Pastor wanted to do a range that would find its place as naturally on a terrace as beside a pool, and may even be used in the interior. Sober and snug in spite of its light-weight metal structure, it comes in black and white finishes.

OUT LINE

Design Agency: EXPORMIM

A clean, enveloping contour gives 'Out Line' a contrasted edge that shifts between armchair, unit chair and even bar stool. This is a future basic for the interior developed by Nieves Contreras, a young designer now working in France who has collaborated with Gabriele Pezzini, eliumstudio and Pascal Mourgue. She projects 'Out Line' as an elegant and simple seating solution—metal structure with stretched polyester. An archetype folded to offer more comfort. 'Out line' comes in two versions: polyester or 6mm steel tube.

Pixi

Design Agency: EXPORMIM

Onyx

A Belgian architect with a genuine concern about the environment and past experience with eco-friendly buildings, Gaëtan Van de Wyer is graduated from l'Instituto CNOFA di Fiora, Bologna (Italy) and from l'Institut Supérieur d'Architecture de la Cambre, Brussels (Belgium).

Along with his architectural projects in Europe, Gaëtan Van de Wyer designs the entire Onyx collection.

Selecting natural materials found locally, he transforms them to create furnitures that don't leave anyone unmoved.

Oxbow

Design Agency: Onyx
Designer: Gaëtan Van de Wyer

This asymmetric sofa is made of a single folded strip whose both ends seem to be unattached. The back of this 3 seater sofa gently floats and is delicately laid on the basis.

Klein

Design Agency: Onyx
Designer: Gaëtan Van de Wyer

Going a step forward in his exploration of abstract wonders, Gaëtan Van de Wyer started from the Klein bottle, a mathematical figure, with no distinct 'outer' or 'inner' surface. The 'mouth' of this chair is prolonged into the 'tail' in an endless surface. But, beside this scientific origin, Klein is firstly a comfortable and amazing chair, and you don't need to know anything about mathematics to enjoy its unusual shape.

Moebius

Design Agency: Onyx
Designer: Gaëtan Van de Wyer

Created under the slogan 'Double armchair, twice the pleasure', Moebius is an endless armchair inspired by the Moebius strip, symbol of infinity ∞. This piece of furniture is designed to encourage users to sit facing each other, prompting conversation or a romantic liaison. Unlike regular sofas, the reverse seating means the gaze of the other person can be easily met and held. Interesting also when it comes to partying and guests have to come up with 'special' sitting positions when talking to each other.

Kenneth Cobonpue

In 1996, Kenneth Cobonpue began to manage Interior Crafts of the Islands, Inc. (ICI). ICI is a furniture design and manufacturing company founded in 1972 by Kenneth's mother Betty Cobonpue, a designer famous for creating new techniques in working with rattan. Discovering that modern design could have a new face using natural fibers and materials as the medium, he slowly created the pieces, which form the collection today. Integrating locally sourced organic materials with innovative hand-made production techniques offered an alternative to the Western definition of modern design.

Noodle

Design Agency: Kenneth Cobonpue

Made of rattan loops resembling spilt noodles or a random doodling, Noodle displays the excitement in the accidents that drive the imagination. The collection succeeds in its spontaneity matched only by the flexible and versatile functions of the material creating pieces that celebrate the occasion of chance.

Lolah

Design Agency: Kenneth Cobonpue

Using traditional boat-building techniques in bending rattan, Lolah integrates old technology with modern design. The collection showcases the flexibility of its material, fastening rattan on a frame to create a variety of sensuous forms.

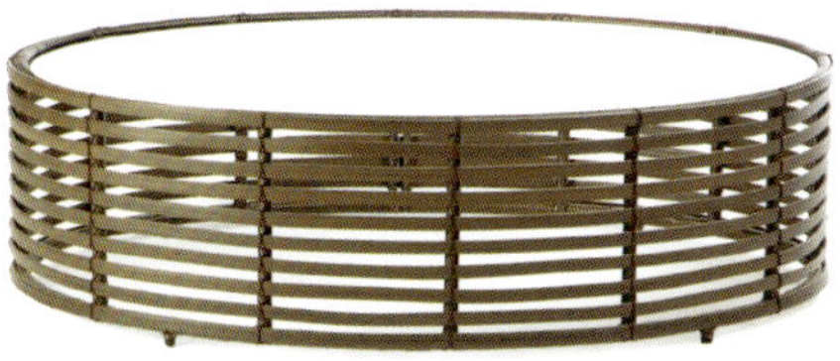

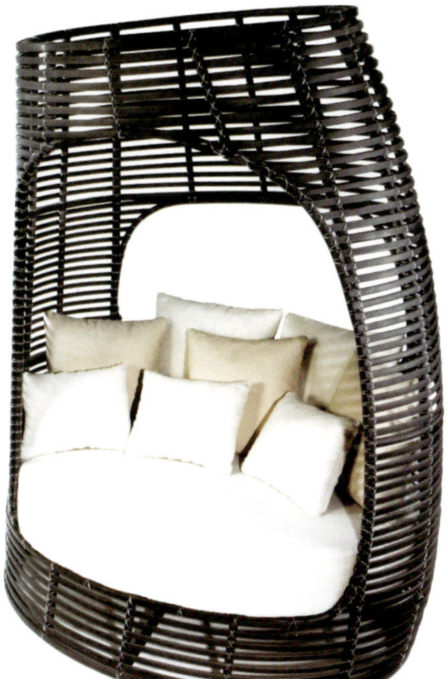

BALOU

Design Agency: Kenneth Cobonpue

Balou exudes comfort at first glance. The collection exhibits an amiable roundness that assures a pleasant seating experience. The airy weave balances the full forms with a light and breezy look.

La luna

Design Agency: Kenneth Cobonpue

La Luna resembles the quiet sophistication of the moon, brought closer to the more tactile senses. The collection is defined by a roundness that typifies its muse and is subtly woven with two woven patterns. A work of precise craftsmanship and design, La Luna can effortlessly float in any space.

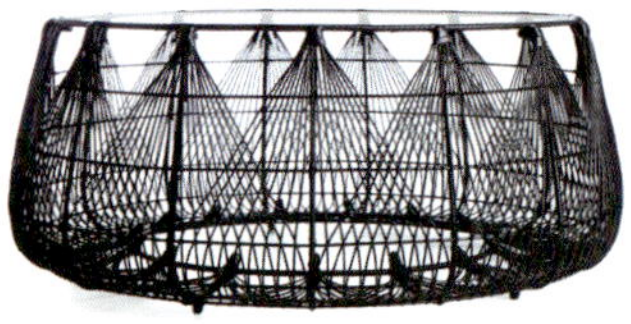

HAGIA

Design Agency: Kenneth Cobonpue

Your search for shade ends with the Hagia. Bask in its royalty and luxury as its gothic temperament echoes a lofty haven.
Skillfully crafted using synthetic fiber, the Hagia is intricately hand-woven on a lightweight aluminum frame. It's airy and transparent personality stays true for its part in your outdoor enclave.

VOYAGE

Design Agency: Kenneth Cobonpue

Voyage cocoons the sleeper into the realm of dreams. Reminiscent of ancient reed boats, the Voyage collection takes the dreamer into journey, a boat ride into the world of the imagination.

I'M A BED

Design Agency: Kenneth Cobonpue

Inspired by ancient oriental temples, The Ima is a true sanctuary. Traditional abaca rope is then intricately woven into a canopy creating the perfect holistic experience. Great with other oriental inspired designs.

TILT

Design Agency: Kenneth Cobonpue

Tilt rests on a very simple idea—the tilt. Embracing the reality of recline as a pose of ease, Tilt succeeds not only in its ergonomic functions but as well as its visual appeal. The Tilt collection inclines to a very comfortable and stylish direction.

Matilda

Design Agency: Kenneth Cobonpue

Matilda communicates richness in its robust forms and invigorating surfaces. The rattan weave complements the hearty shapes that provide a sense of fullness to the collection. Resembling molded baskets, the collection features rattan's beauty and durability through a variety of processes.

Stitches

Design Agency: Kenneth Cobonpue

Multiple forms can be created with these tables that are wrapped in plain or stamped leather, then stitched with polycotton thread. Metal Casters are bottom.

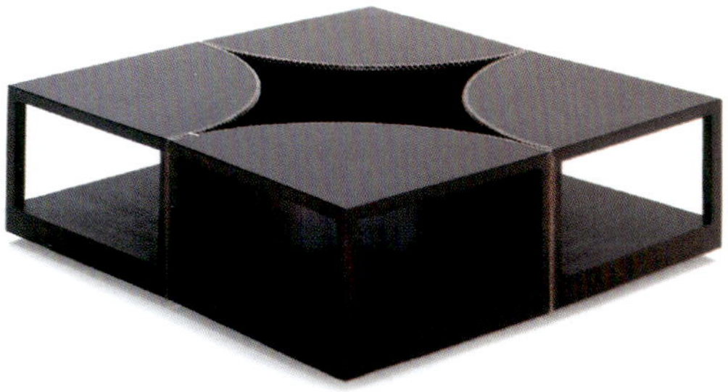

Kawayan

Design Agency: Kenneth Cobonpue

Named after the local term of the material, the Kawayan collection features the perennial Bamboo. Showcasing its natural beauty, the distinctive nodes and natural color are retained, at the same time showing its strength and versatility in is structural and aesthetic functions.

Oasis

Design Agency: Kenneth Cobonpue

Sink into an oasis of comfort. The tulip-inspired Oasis is made of polyethylene strips handwoven on an aluminum frame. Several pillows make this collection extremely comfortable.

Dragnet

Design Agency: Kenneth Cobonpue

Made of fabric twisted and wrapped on steel, the collection's dynamic forms and patterns evoke an excitement that stimulate the senses. With its vibrant color choices, Dragnet instinctively adds a contemporary chic flavor, leaving any space caught in its net of style.

Nobu

Design Agency: Kenneth Cobonpue

A collection of minimalist tables laminated in different patterns of leather, Nobu's sophistication lies in its simplicity.

Cho Hyung Suk

Cho Hyung Suk is interested in the furniture structure than external form. He thinks new structure at first, and if he designed new structure, an external form becomes naturally the design along structure. A design to have taken out structure isn't anything. Cho Hyung Suk wants 'Good design' to give smile to everyone. It isn't important any form and any idea. He wishes everyone become happy because of design.

STUD CHAIR

Designer: Cho Hyung Suk
Dimensions: 55cm x 56cm x 67cm
Materials: steel pl, bending wood

The 'Stud' means a small button like earrings. The 'Stud chair & table' is wearing four small earrings. I want to design the chair composed by only one module.
The 'Stud chair & table' solves this problem. Its back, sitting plate of chair and table are only one bending module. So, it is maded easily and its cost is more cheaper than 2 modules or 3 modules design.

CURVE FRAME SOFA

Designer: Cho Hyung Suk
Dimensions: 198cm x 64cm x 94cm
Materials: sts steel pipe, fabric

The Curve Frame sofa consists of one lounge sofa, side table and stool. A sofa and stool have three cushions. The frames have smooth curve and hold each cushions. It solves the structure each of them.

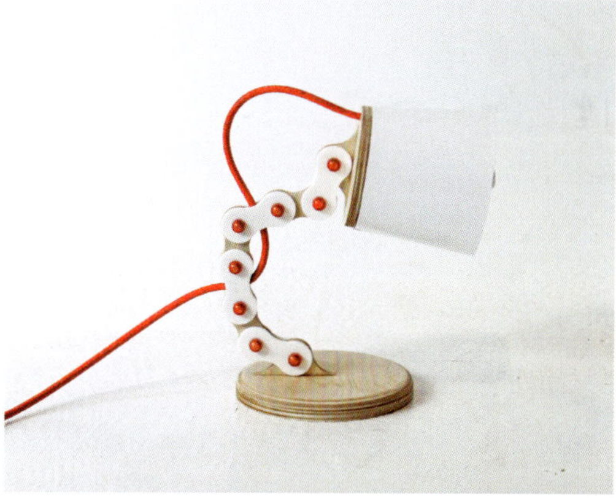

B-chain lamp

Designer: Cho Hyung Suk
Dimensions: 23cm x 17cm x 22cm
Materials: birchplywood, Stone, hardware

The 'B-chain lamp' can have most various shapes than any lamp.This is motivated from the bicycle chain. So, its 8 joints can shine wherever you want to light. It is consisted of four type modules. It is possible whatever length and shape you want to make.

Modernatique

Designer: Cho Hyung Suk
Dimensions: 52cm x 60cm x 60cm
Materials: walnut, ash

In general, furniture is divided in two styles, Modern and Antique. These two styles are very different in appearance, so it is very different from each attraction. Modernatique chair has each element of these two styles. Beautiful curves of antique and clean lines of modern design are in proper harmony.

AAN-UIT

The AAN-UIT collection distinguishes itself with its enticing lucid lamp designs by using today's shapes and materials. The label is built on the philosophy that 'light' should be explored to the max. 'Since an environment is changed by light, the designs create specific different experiences whenever they're turned on or off. That's why lighting fascinates so much'.

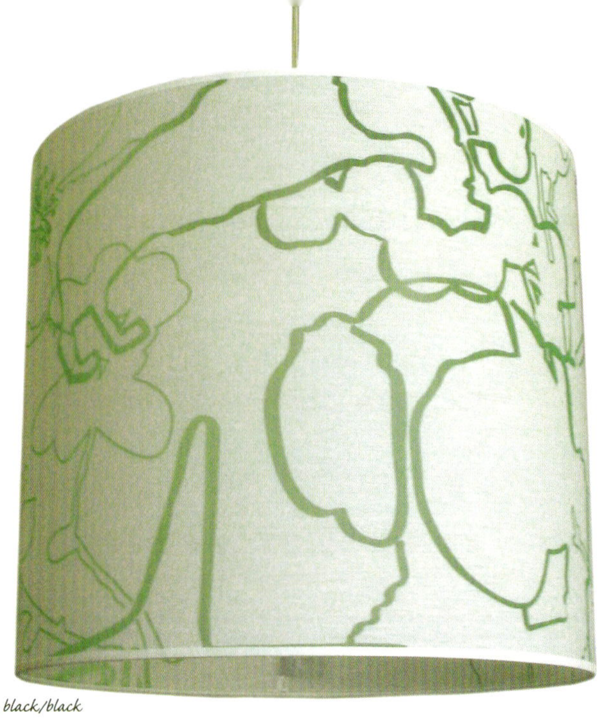

Fantasy

Design Agency: AAN-UIT
Colors: white/green, white/white, black/black
Dimensions: 40cm x 38cm

Moon night

Design Agency: AAN-UIT
Dimensions: 75cm x 140cm/ Large: 100cm x 195cm

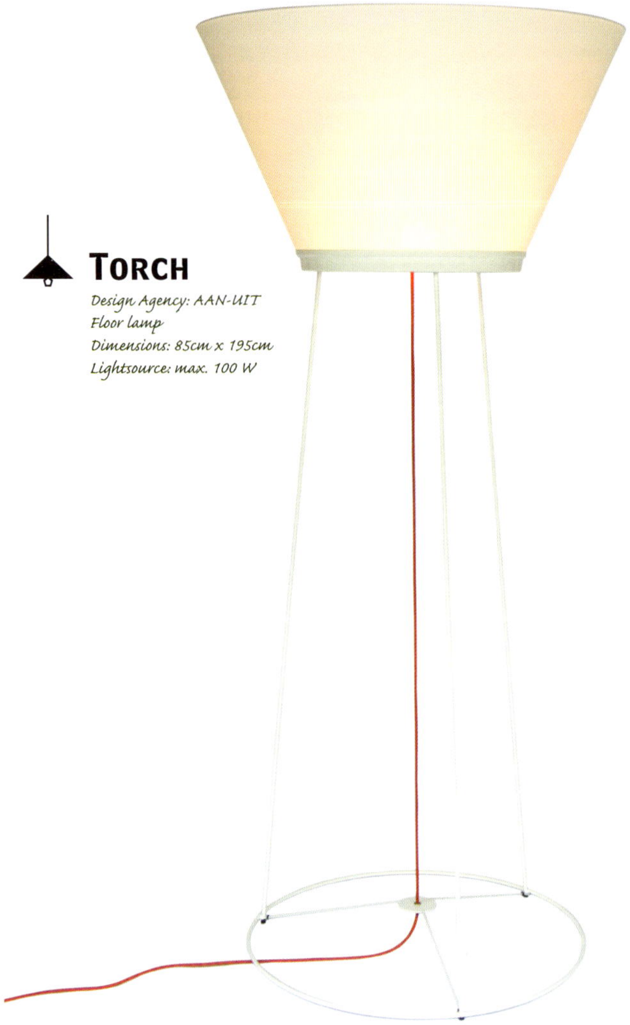

Torch

Design Agency: AAN-UIT
Floor lamp
Dimensions: 85cm x 195cm
Lightsource: max. 100 W

Roots

Design Agency: AAN-UIT

Dimensions: 60cm x 155cm

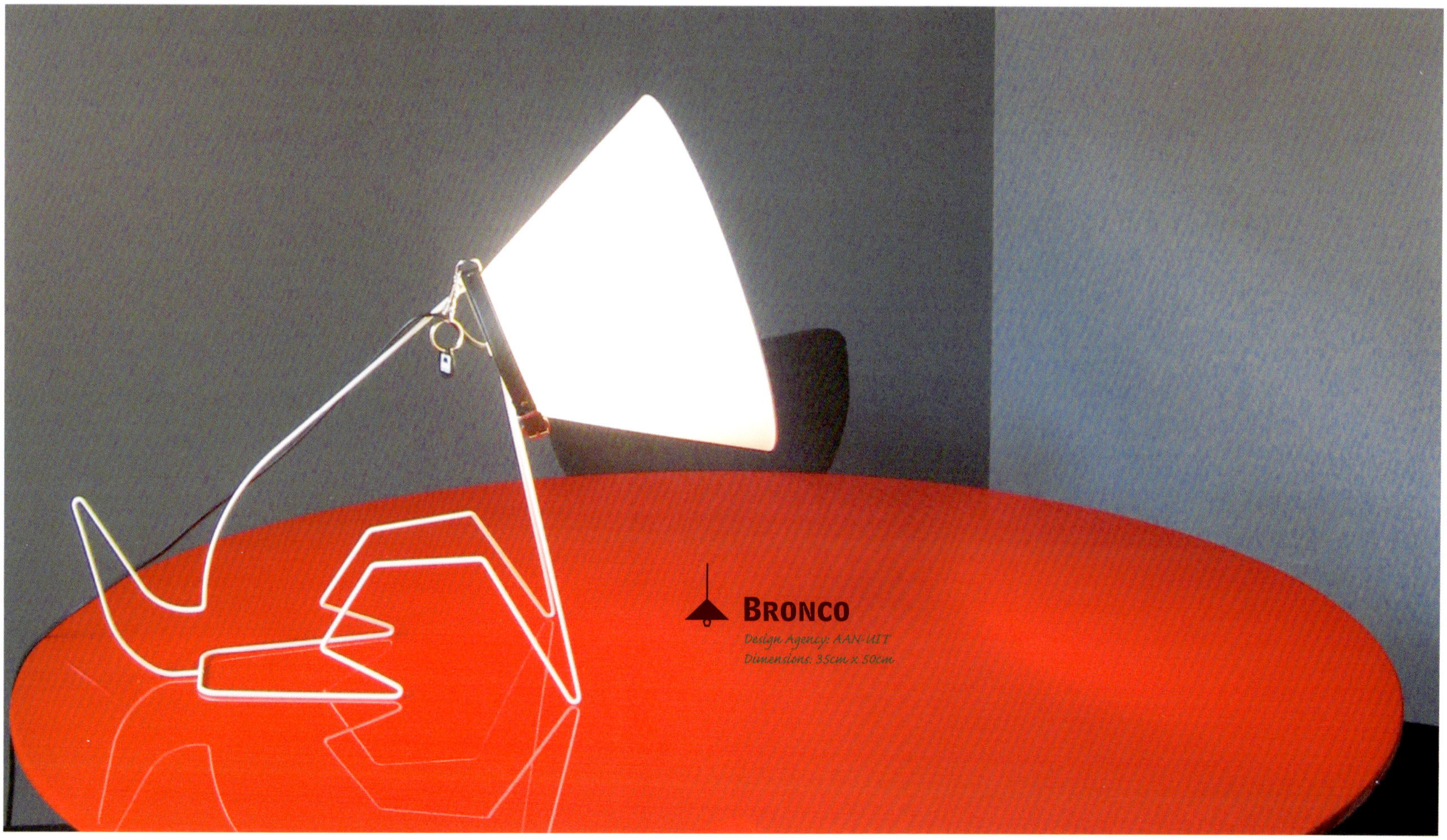

Bronco

Design Agency: AAN-UIT

Dimensions: 35cm x 50cm

Walter Knoll

Walter Knoll is one of the leading international manufacturers of fine upholstered furniture and office facilities. With designs both classic and modern, with peerless comfort and great durability, our products create living spaces.
Walter Knoll is known as the oldest upholstery brand in Germany. Founded in 1865, the company continues to make its mark in the history of furniture; it has created classic pieces and icons of the avant-garde and is a by-word throughout the world for modern design.
Inviting and prestigious as well as internationally comprehensible in design, Walter Knoll furniture makes the desired quality of life a reality. Together with renowned architects and designers the company creates furniture and furnishes the interiors of buildings worldwide.

Cuoio

Designer: EOOS

The logic of minimalism determines the shape and materials of the Cuoio Chair. Leather and steel are the components of surface and framework. Minimalist and timeless. The fact that the chair is separable and recyclable ensures it is a miracle of sustainability. The slits in the seat and back lend the chair its typical shape. The flexible back ensures comfort. Shaped for comfortable sitting. The Cuoio Chair comes both with and without armrests, entirely in black or in natural shades of brown.

MYchair

Design Agency: UNStudio/Ben van Berkel

A new slant on things. MYchair is a sculpture in space. Sweeping and expressive. Impressive from all angles. Slanting, rounded and straight areas give way to each other. The base, seat and back of the lounge chair form a powerful composition. The upholstery traces convex and concave shapes. Intense colors create dynamic duets. The lounge chair is appealing and comfortable, and invites you to read, dream and relax.

TOGETHER

Designer: EOOS

In pursuit of new solutions for sitting together at table, the Austrian design team EOOS has rediscovered the corner seat—an element which for centuries has repeatedly been a success. It is the meeting point within a house, the place for discussions, for eating together, the stage for the rituals of just being together. The new interpretation is the Together model. A sculpture of architectural clarity with the seating comfort of good upholstered furniture. An impressive appearance perfectly defines the meeting place even as a stand-alone item—both at home and at work.

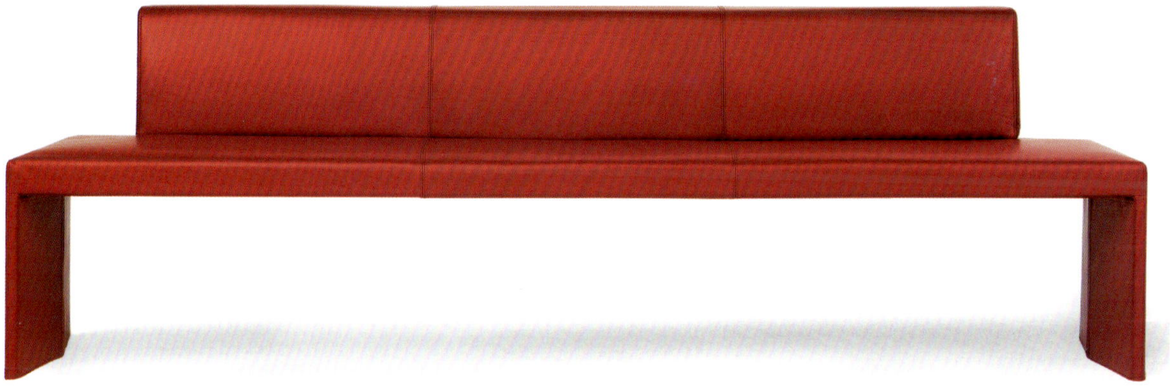

FK

Designer: Preben Fabricius & Jørgen Kastholm

The Devil Wears Prada... and sits on FK. Like Meryl Streep on the screen, aficionados worldwide worship its unmistakable appearance: FK—the unique form, the clear contours, the perfect lines. The bucket seat is an icon of minimalism in design. This new interpretation of a classic emphasises the outer shell. Chic and clean, shining and stylish. The leather cover on top, inviting and familiar. The new executive chair—not only for diabolic editors-in-chief. Devilishly good.

April Hannah

April Hannah is an artist and designer working in Brooklyn, New York. She received her MFA from the School of Visual Arts in 2001. Hannah's first foray into furniture design was realized in May of 2010, with the debut of the Tree Table Collection at BKLYN Designs, where she received BKLYN Designs' Inhabitat Editor's Choice Green Awards for Best Kids Design and Best Newcomer. In December 2010, she was awarded a residency at MAD Museum (Museum of Arts and Design) in New York City. Hannah's work has been shown in both solo and group national exhibitions and has been reviewed by numerous magazines, newspapers and design blogs.

THE TREE TABLE COLLECTION

Designer: April Hannah

Inspired by the networks and patterns of nature the eco-friendly collection exhibits a modern, simple yet sophisticated, playful aesthetic that engages both children and adults. The pieces are geared for children ages 5-10 and also include an adult-sized Tree Desk.

Lincoln Kayiwa

Lincoln Kayiwa graduated as a Master of Arts from the University of Art and Design Helsinki UIAH in 2007 upon which he started his own company—KAYIWA.
He was born in an architectural family in Kampala, Uganda in 1979 and currently lives in Helsinki, Finland.

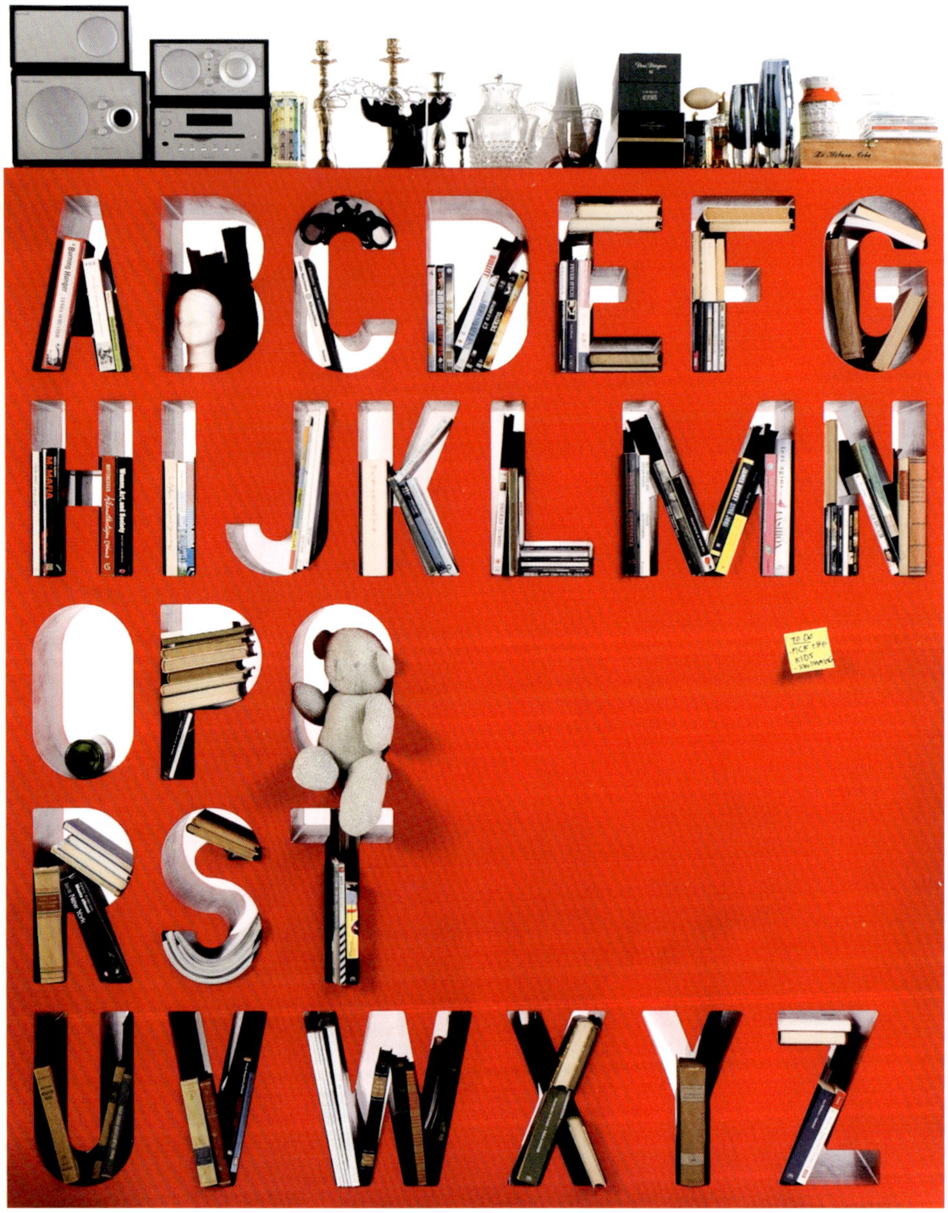

AAKKOSET SHELF

Designer: Lincoln Kayiwa
Material: medium density fiberboard

You may use both AAKKOSET's obverse and reverse to hold books, CDs, magazines et cetera. As a room divider, you may also choose to leave it empty. It is available in black, blue, green, orange, red, violet, white and yellow.

Nico Kläber

Nico Kläber is a German industrial designer born in 1981. He studied design at Köln International School of Design in Germany and the Hong Kong Polytechnic University. He successfully took part in several design competitions such as Electrolux DesignLab 2009. His work has been published in magazines and exhibited on shows and fairs all over the world.

Lounge Chair

Designer: Nico Kläber
Photography: Astrid Wolff

This huge lounge chair is a piece of furniture for hotel rooms, lounges and living rooms. Its oversize and the green felt makes it a focal point in any interior. The user is free to sit on it whatever he wants, except normal. One could say this chair makes hanging out socially acceptable.

Konsolen

Design Agency: Duravit

The Fogo consoles just 36cm instead of 55cm deep open up completely new possibilities for planning smaller bathrooms. In combination with the matching console basins or semi-recessed washbasins, they are suitable for comfortable, narrow washing areas. In widths from 900mm to 1800mm with either one or two drawers, they offer maximum variety.

Duravit

As a tradition-rich company with headquarters in Hornberg, Germany, Duravit AG has had a decisive influence on the development of the sanitary ware branch. Much has changed since the company's foundation in 1817. The initial production of earthenware crockery gave way to the manufacture of sanitary ceramics. In the meantime, Duravit has become a pioneer when it comes to design and innovation in the bathroom segment.

Bold design has become a badge of the Duravit brand. The innovative forms and exciting developments have been crowned with numerous international awards and chalked up outstanding market successes. Shapes that represented unique innovations at the time of their introduction by Duravit have become classics – frequently copied, but never matched.

Studio OOOMS

Studio OOOMS is run by Guido Ooms and Karin van Lieshout. Guido Ooms graduated from the Design Academy Eindhoven in 2000. After working as a freelance designer at different companies he started his own studio. Next to running the studio he also teaches design at the School of Arts in Utrecht. Karin van Lieshout graduated from the Design Academy Eindhoven in 2001. She has a degree in teaching (BIK, professional artist in class) and worked in various designcollectives before she joined forces with Guido.
Both designers have their own qualities, each contributing to a part of studio OOOMS. Guido's enthusiasm and playful approach towards almost everything makes sure that the studio has enough new ideas coming in. Karin's down-to-earth attitude makes sure that the good ideas become reality.

GLASSBULB LIGHT

Design Agency: Studio OOOMS

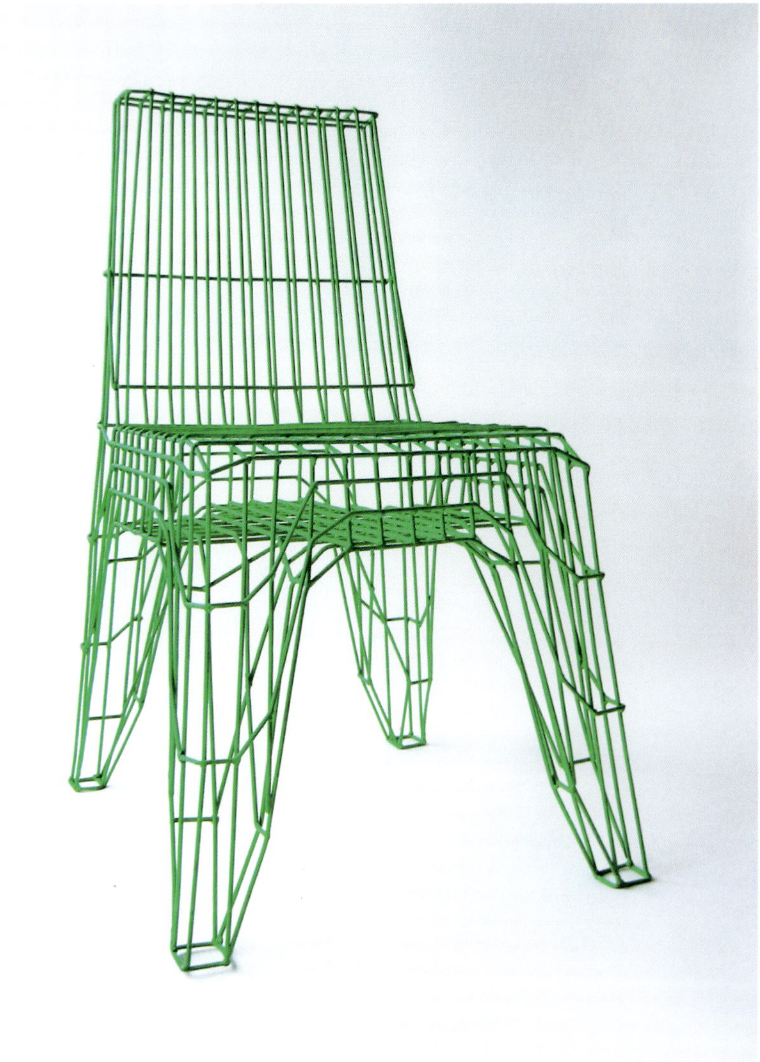

WIRE-FRAME COLLECTION

Design Agency: Studio OOOMS

The Wire-frame collection came into existence because of a great fascination of the wire-frame structures in the 3D modelling phase of the design process. This chair is designed by computer using 3D software and after that it is handmade in real steel wire.

One of the interesting aspects of this chair is that it becomes very transparent; you are able to view the entire chair from one perspective.

LO-RES CHAIR

Design Agency: Studio OOOMS

Schizo vase

Design Agency: Studio OOOMS
Dimensions: 95cm x 50cm

Schizophrenia is a psychiatric diagnosis that describes a mental disorder characterized by abnormalities in the perception or expression of reality. This vase by OOOMS has two faces. Because of the ribbed surface the silhouette of two different shapes can be seen. Is the one shape more real than the other?

Dandelion

Design Agency: Studio OOOMS

Röshults Svenska Hantverk AB

Röshults represent modern Scandinavian design with timeless design language. Their customers value the high quality, solid craftsmanship and tasteful simplicity. Their products create atmosphere and harmony at home, convey a sense of public spaces, transforming the garden into a magical place. Gather friends around a fire in the summer night. Let the garden come into the kitchen. Furnish with light. What do you think happens? They know!

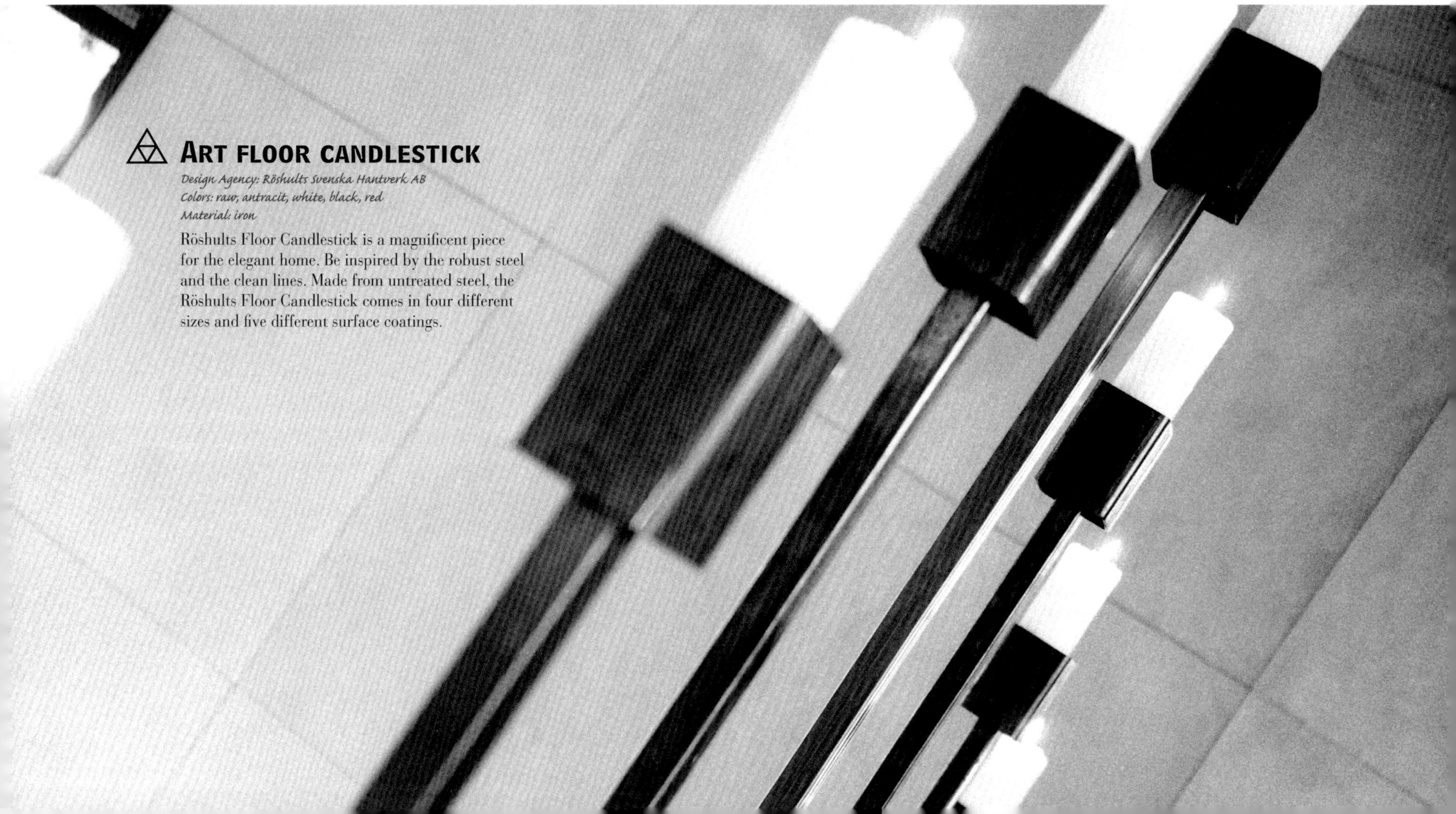

Art Floor Candlestick

Design Agency: Röshults Svenska Hantverk AB
Colors: raw, antracit, white, black, red
Material: iron

Röshults Floor Candlestick is a magnificent piece for the elegant home. Be inspired by the robust steel and the clean lines. Made from untreated steel, the Röshults Floor Candlestick comes in four different sizes and five different surface coatings.

Carl

Design Agency: Röshults Svenska Hantverk AB
Designer: Pierre Sindre

The renowned architect, Pierre Sindre, has designed these modern pots in the Röshults Carl series. This exciting range is synonymous with style and will compliment any home or garden. Made out of iron, the pots are very versatile and durable as the surface is covered with anthracitic grey structure coating which protects and gives a beautiful patina. Röshults Carl is available in two sizes.

Garden bench

Design Agency: Röshults Svenska Hantverk AB
Designer: Broberg & Ridderstråle

The Garden Bench fits the Garden Dinnertable and is made in plated powder-coated steel and the top in Ipe wood. The Garden Bench is available in two different lengths.

Kloss

Design Agency: Röshults Svenska Hantverk AB
Designer:Design Lodge
Material: iron

Kloss is a reversible iron pot, which also can be used as candle holder. Holes for wall mounting. Available in two sizes and different surface treatments.

BBQ GRILL 100

Design Agency: Röshults Svenska Hantverk AB
Designer: Broberg & Ridderstråle
Dimensions: 50cm x 50cm x 80 cm
Materials: powder-coated iron, stainless steel, zinc, chromium

Grill 100 is a stylish charcoal grill equipped with an adjustable airflow for heat regulation, one removable zincked charcoal holder with ash catcher, one chromed cooking grate and yellow chrome plated, powder-coated body. The idiom gets clear inspiration from the world of architecture and the grill fits into a landscape of houses and gardens. The designers are Broberg and Ridderstråle, a duo that works with design and architecture, where the goal is to communicate with the receiver.

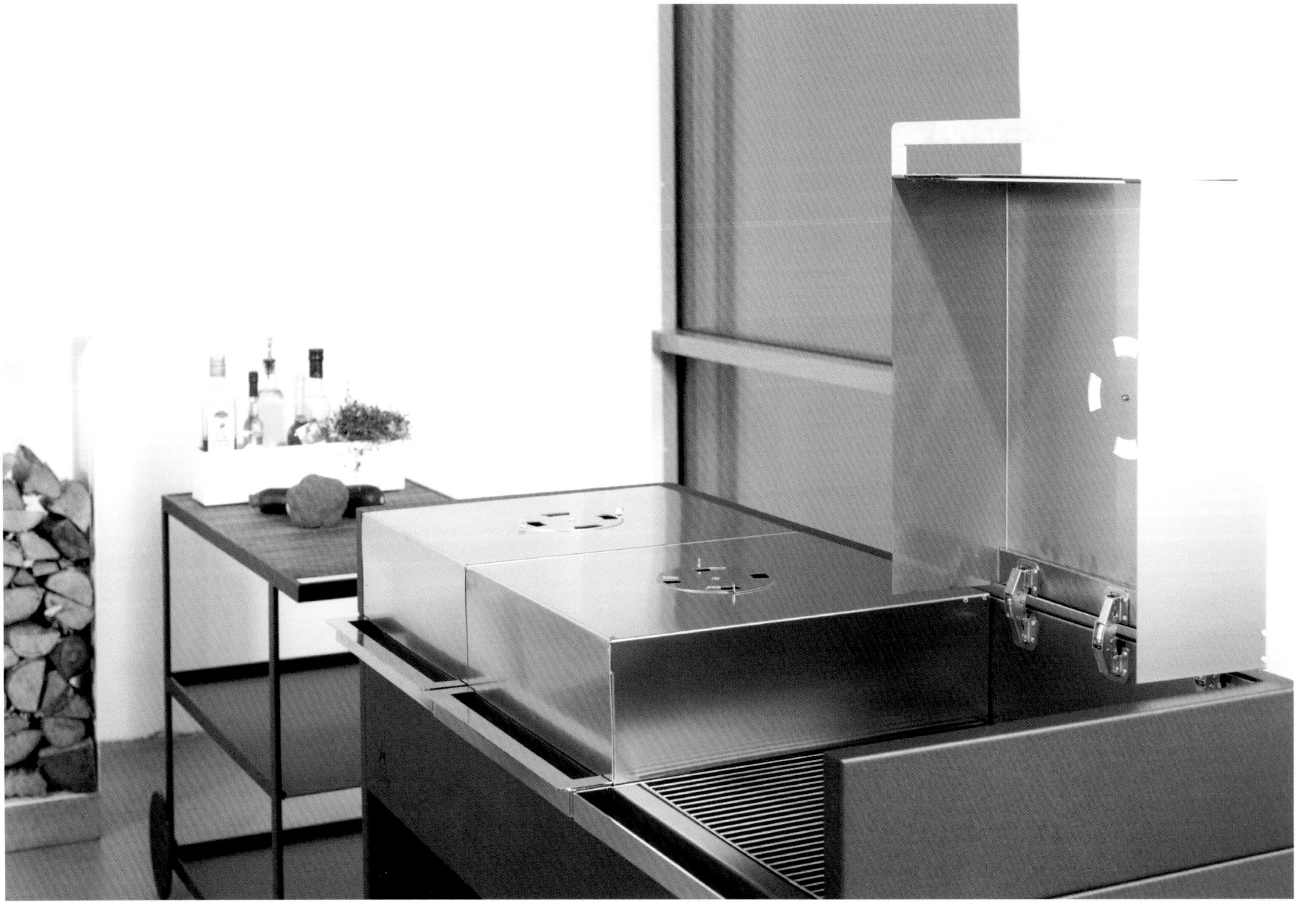

Miso Soup Design

Co-founded MisoSoupDesign in 2004, Minnie Jan and Daisuke Nagatomo dedicate their practice on researching material property, systematic strategy, and re-evaluate functionality. Both being USGBC LEED accredited professionals, they are aware of the direct connection between environment and limited resource, and the concept of their design constantly incorporates new ideas with minimal wasted material. With the aid of digital technology, they began to apply the systematic way of thinking in furniture design, and discover the surprising moment when form meets function. MisoSoupDesign has received numerous design awards, including honorable mention in Biennale Miami 2005, 1st prize in ACADIA competition 2007 and FEIDAD design merit award 2008. Their work has been published globally, including Frame magazine, Tatlin, L'arca, and Interior magazine in Taiwan China.

Kulms chair

Design Agency: MisoSoupDesign
Manufacturer: Lerival LLC
Dimensions: 79.4cm x 47cm x 56cm
Material: CNC-milled lacquered ply

Kulms Chair uses bamboo to achieve its sustainability. The name comes from 'culms' the stem of plants to describe its materiality. Its design attains stackability.

WALTZ LOUNGE

Design: Daisuke Nagatomo & Minnie Jan / MisoSoupDesign
Manufacturer: OFS

Waltz Lounge and Ottoman, the winning entry of OFS the Hotseat 3 competition designed by New York City based MisoSoupDesign / Daisuke Nagatomo and Minnie Jan, has been unveiled its beautiful curvature in OFS Showroom at NEOCON 2009. The curving sheet of molded wood leaves space for stashing an ottoman. Waltz is Green Guard product and also available with FSC Certified Wood.

K WORKSTATION

Design Agency: MisoSoupDesign

K Workstation incorporates Bamboo laminated plywood for creating efficient space managing working space. The folding curved shape creates shelving and desk space within one working unit. K Workstation could be manufactured each segments for easy fabrication process and transportation, yet it could be configured with multiple possibilities. The bamboo, rapidly renewable material, brings this working unit sustainability and elegance of curvature.

Bakery studio

Bakery studio was founded in September 2009 by Gil Sheffi, Ran Amitai and Gilli Kuchik, the three met and became friends during their studies at the Bezalel Academy of Art and Design in Jerusalem Israel .
The designers believe in hands on experimentation with materials and technologies and that in order to achieve innovation, you have to surrender to an unknown journey. Materials are perfect partners for this journey because they surprise you. Sometimes the material responds to the designer wishes and sometimes the designer responds to the material wishes.

Elastic wood

Design Agency: Bakery studio

The elastic wood project focuses on joinery, seeking to explore what happens when they become flexible. The project is inspired by the common process of overmolding, two materials are used to complete and enhance each material characteristics; The wood is used for its compatibility for home furniture which is welcoming to touch and transmit a warm aesthetics. The elastic material is used for the propose of transforming the 2D spread into a 3D shape. The combination of the two materials adds the aspect of flexibility into the usually rigid furniture produced in wood.

Industrial upholstery

Design Agency: Bakery studio

Industrial upholstery is a project which explores new ways of upholstering furniture. During the research stage of the project, it was apparent that upholstery as a discipline is based on a craftsmanship logic which stayed fairly the same through upholstery history. The Industrial Upholstery project seeked to create a new upholstering technique which is based on an industrialized logic, and by doing so creating a new aesthetics of upholstery.

Nature of material

Design Agency: Bakery studio

The idea of the project is to combine straight origami like folds with natural folds. All of the objects are created by planed and exact folds that create side effects of stretching and bending in the material. In other words, by folding the legs of the objects the surface above reacts naturally to the folding act.

Büro für Form

The Büro für Form design studio was founded by Constantin Wortmann 1999 in Munich, Germany. The main focus of its work is on product, interior and lighting design for renowned, upmarket, international manufacturers like Georg Jensen, Umbra, Kundalini or Next. The red thread of the design is the well-balanced combination of organic shape and geometric elements, sometimes with a bit of humor, creating a particular modern style.

Mission statement: Products need more then perfect function and ergonomics, they need some poetry! Works of Constantin Wortmann have been featured in many publications, shown on international exhibitions and garnered numerous international design prizes.

Atomium

Design Agency: Büro für Form

Inspired by the famous Atomium Sculpture in Brussels, this sculptural light has six endings with integrated light. There are versions as floor, wall and pendants available.

NLC

Design Agency: Büro für Form

At a first glance, NLC perplexes with its positively striking and yet strangely puzzling appearance. Its special appeal lies in the seeming conflict between the chaotic and the organic. NLC draws its inspiration as much from the head of Medusa as from the finely trellised branches of a treetop. One of the reasons for the luminaire's attractiveness certainly stems from the fact that the design invisibly incorporates the light source. The white cover hides the light fittings and is easily adjusted to create the desired amount of illumination. Dimmable LEDs offer an additional option of controlling the level of brightness. Due to the E14 fittings NLC can be used either with LEDs or standard energy-saving lamps

DNA

Design Agency: Büro für Form

Like an organically growing structure, the complex form develops out of a suprisingly simple element. White, pink, chrome and gold plastic modules—each with three light bulbs—can be added to each other, so the lighting object can vary in size, from a single format up to a luster. You have the choice between seven different assembled pendant versions, a table version, a wall/ceiling version or, alternatively, a customized solution according to your needs.

Liquid light

Design Agency: Büro für Form

Liquid Light presents light in a fluid form. The different shapes of Liquid Light offer a warm and diffuse light for any space. Illuminated drops float weightlessly in the air, run down a wall, build a soft rain of light or melt out of the ceiling like a viscid fluid. This award-winning series of lamps visually enhances every type of architecture.

FLAPFLAP°10

Design Agency: Büro für Form

FlapFlap°10 playfully defies the laws of gravity. The luminaire appears to hover freely in the air. As if lifted by phantom hands the invisible reinforced cable rises up from the ground to support the lampshade. The first version of FlapFlap designed by Constantin Wortmann and produced by Next home collection dates back to the year 2000. On the occasion of Next's tenth anniversary Wortmann has created this redesigned version. The outcome is a brilliant, sparkling FlapFlap°10. In accordance with the spirit of the time, the luminaire is available with LED bulbs. However, the E14 fittings are also appropriate for use with energy-saving lamps.

VELA

Design Agency: Büro für Form

VELA can be seen as an abstract part of broad organic network or as a natural lace in expansion. The dynamic shape with its amorphous appearance links the objects together: a magazine rack and a jewelry tree. It is sculptural yet functional.

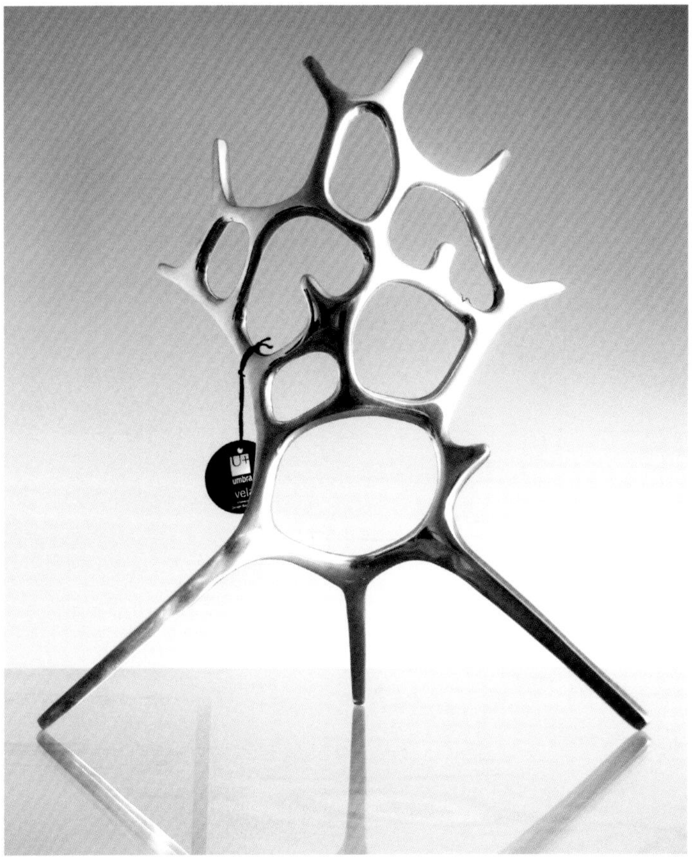

Straight Line Designs

Straight Line Designs is a one-of-a-kind workshop that has been operating out of Vancouver, British Columbia for the past 25 years. In addition to installations, sculptures and private commissions, designer Judson Beaumont and his staff of eight full-time craftspeople have designed and constructed a variety of custom-built furniture and projects for public institutions and children's exhibitions throughout North America and abroad. Focused on quality and custom design, Judson's studio stays far away from mass production. Our workshop's building capacity is restricted but Judson's designing ability is not!

OOPS! CABINET

Design Agency: Straight Line Designs

'What if' is one of the questions Jud always applies to furniture. What if my furniture melted? What if it crashed into something? Or in this case, what if something crazy happened like a cannonball was shot through my dresser? The results are always fun and sure to make people think about their furniture a bit differently.

CRASH TABLES

Design Agency: Straight Line Designs

Wooden rock cab

Design Agency: Straight Line Designs

Small squiddy table

Design Agency: Straight Line Designs

Stanislav Katz

Born 1983 in Riga, Latvia. First took interested in computer graphics in 1995. In 2001 started working with local interior designers as an assistant and a CAD artist, learning about industrial and interior design. In 2005 started working on personal projects including furniture and home accessory concepts as well as interior design concepts. In 2008 took part in Baltic Furniture Manufacturers Design Award 2008, won a prize in Unique Design category for 4 Vase Cabinet. In 2008 co-founded the company Katz which specializes in interior and industrial design, currently working there as an art director.

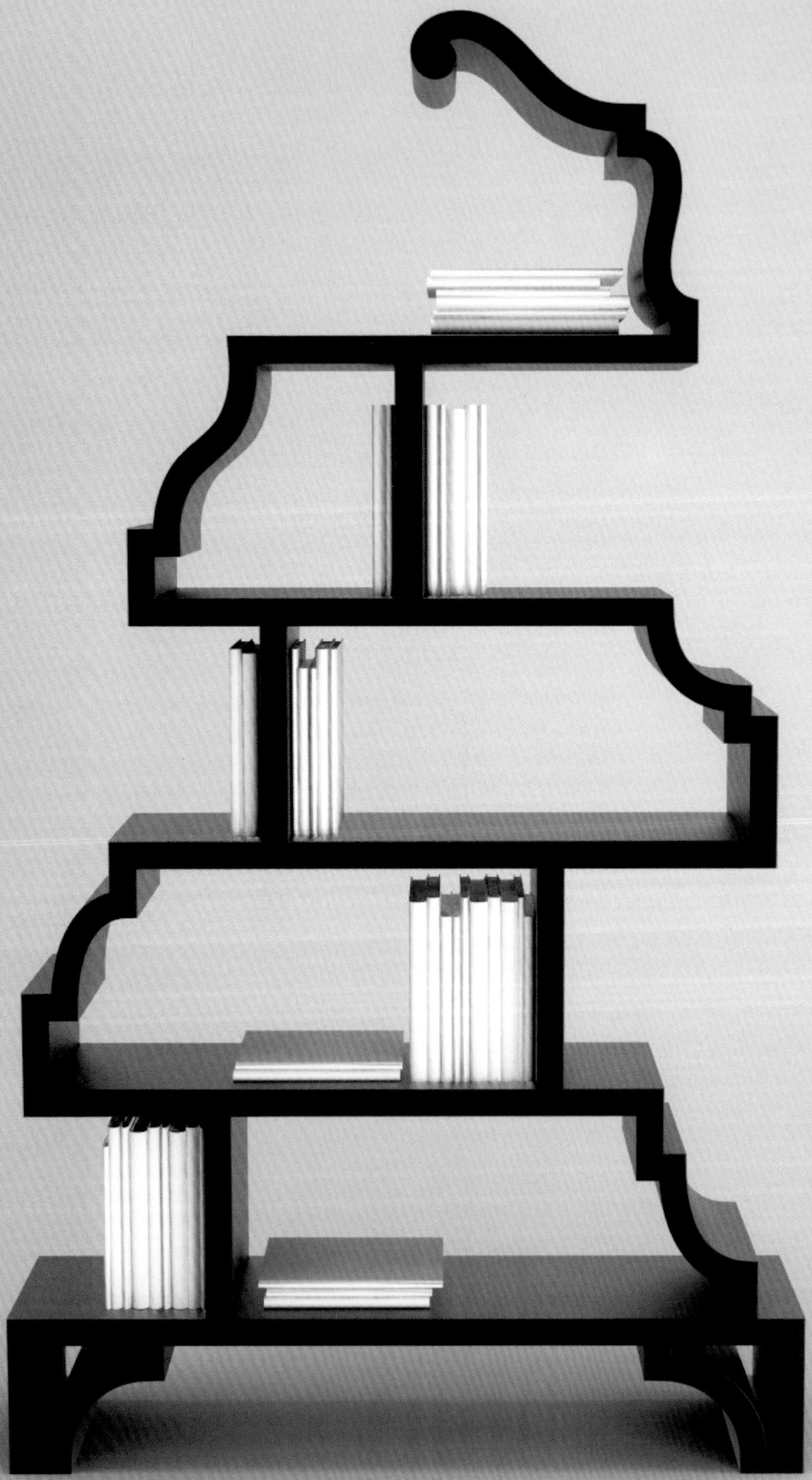

DECAY SHELVES

Designer: Stanislav Katz
Dimensions: 213cm x 102cm x 30 cm
Material: painted MDF

This stylish neo-baroque bookshelf adds a hint of glamorous decadence to your apartment.Handmade in EU.

Drawer's Chest

Designer: Stanislav Katz
Dimensions: 130cm x 97cm x 50 cm
Material: painted MDF

Drawer's Chest is a fusion of contemporary and classic motives, bringing together functionality and style. It has 6 single narrow drawers for smaller things and 3 wide double drawers for bigger stuff. Handles are custom made from chromed metal.

Console bookshelf

Designer: Stanislav Katz
Materials: painted MDF, polyurethane
Dimensions: 215cm x 110cm x 50 cm

This unique baroque-inspired piece of furniture offers you a great way to relax being surrounded by your favourite books.It consists of two parts sealed together—the inside is made out of soft yet springy polyurethane which allows you to sit or lay down on it. The outside part is solid, handcrafted and painted MDF.

Studio Hanna & Seo

Jeonghwa Seo(1982) graduated in metal art & design at Hongik University in Seoul and Hanna Chung(1981) graduated in visual communication design at Kyoungki University in Suwon are Studio Hanna & Seo's two Korean designers based in Eindhoven, The Netherlands and Seoul, South Korea.
The collaboration started during their Master course in IM of Design Academy Eindhoven. After Jeonghwa Seo has received master degree with his project 'the geography of thought' in 2010 in collaboration with Hanna Chung , they set up a studio based in Eindhoven and South Korea. Combining multidisciplinary intellectual background gives broad perspective on delivering their thoughts into context of design.

Ripple effect tea table

Design Agency: Studio Hanna & Seo
Designer: Jeonghwa Seo, Hanna Chung
Materials: ash, plywood, earthen ware, pure water
Dimensions: 120cm x 50cm x 35cm

There is a belief by Asian people that small changes in an individual's life can have an impact and have a certain effect to the whole community, which in a psychological term is called a 'ripple effect'.
Hailed as the 'Ripple effect tea table,' the table blends Eastern ritual with the social relationships. Each time you place the cup on the table surface, the ripples can be seen speeding up on the surface of the body.

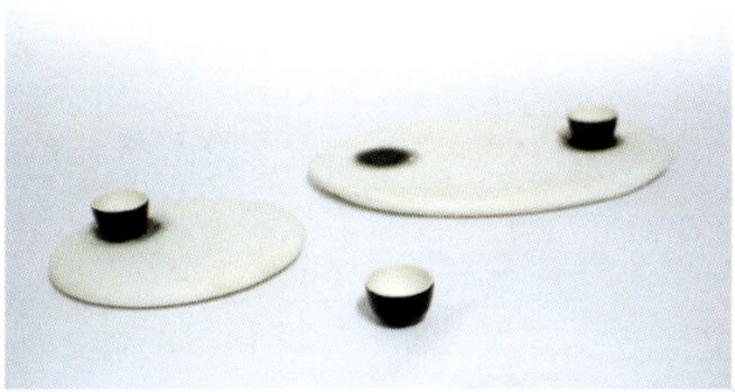
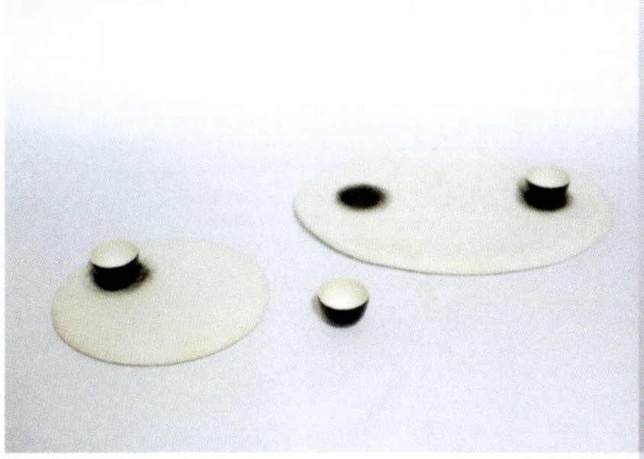

VOID TEA SET

Design Agency: Studio Hanna & Seo
Dimensions: 30cm/50cm x 27cm x 7.5cm

The eastern philosophies that guide the Korean culture see no boundary between man and nature, existence and non-existence, focusing on the void and transition between them. The tea set offers a touch of relaxation and meditation, in the midst of busy daily life.

Emiliana Design Studio

Ana Mir and Emili Padrós are MA ID Master of Arts in Industrial Design in the Central Saint Martins School of Arts and Design, London. In 1996 Ana and Emili founded Emiliana Design Studio based in Barcelona (Spain), working together on different commissions and at the same time developing individual research projects. Emiliana Design Studio combines product design, furniture, lighting, design spaces and exhibitions. Several of their designs are submitted to the permanent collection of Indianapolis Museum of Art, Fons Nationale d'Art Contemporain of France and Museu d'Arts Decoratives of Barcelona. They participated at MoMA (Museum of Modern Art of New York) in 'Safe: Design Takes on Risk' and 'Design and the Elastic Mind'. The vision of Emiliana Design Studio begins from the experimentation and the prospective new scenes, where the objective is that the relationship between objects, spaces and users are the most emotional, participatory and creative potential.

Kiosk

Design Agency: Emiliana Design Studio

Naoshima

Designer: Ana Mir+Emili Padrós, Emiliana Design Studio
Manufacturer: ZIRU (Valencia, Spain)

Emiliana Design Studio wanted to design a stool in which all the parts were related to each other, a stool that could be understood as a continuous object, and not as four legs added to a seat. They started making scale models, trying out structures, joints ... and they imagined that the folds they were trialling with paper would end up being made in curved plywood. They put a lot of effort into studying the joints between the seat and the legs, both at a structural level, but also from the aesthetic point of view; they started to like the idea that the whole stool was to be produced in wood, and that the legs could be seen as prolongations of the seat itself.

KETTAL

Design Agency: Emiliana Design Studio

Serien. Lighting

25 years on and Serien. Lighting is a firm fixture in the world of technically discerning luminaires with a quality design. Luminaire design from Germany is, among other things thanks to the duo, now regarded worldwide as a guarantee of quality. And the market has duly rewarded the two designers' standards in terms of the quality and manufacture of their ingenious lighting solutions.

They were technical and functional, but also elegant in thrust and precisely formulated solutions. Each of them is the result of closely addressing the respective design, technological and lighting issues. This was the policy for all the products that Serien.Lighting has made in the intervening years. Over the 25 years it has repeatedly caught the eye and stimulated new luminaire design.

Zoom

Design Agency: Serien.Lighting
Designer: Floyd Paxton

Zoom can transform itself: it is a circular concertina grille, enabling the diameter of the luminaire to be adjusted ad infinitum. The smallest setting has a diameter of just 20 centimetres, while opened up wide it measures 1.30 meters.

Zoom XL is twice as big as Zoom: closed the circular luminaire measures 40 centimetres in diameter, and completely extended 2.60 metres. If required, both versions can be adapted to the surroundings in terms of both shape and lighting effect. Given its dimensions, Zoom XL lends itself particularly well to large, high rooms. The smallest version—Zoom—comes equipped with halogen bulbs. Zoom XL is equipped with screw sockets and can therefore accommodate a variety of bulbs, colored for example.

Poppy

Design Agency: Serien.Lighting
Designer: Ulrich Beckert, Georg-Soanca Pollack, Peter Thammer

Switch on Poppy and the shade opens up like a flower. This is made possible by the fine bi-metal leaves, which adopt a predetermined position through the warmth of the lamps. Poppy – the name says it all: its bud, the goblet-like shade made of mouth-blown glass that comes in red, ceramic, and dark violet is slightly transparent when lit. Its stalk, made of flexible tube in black, or beige is fabric-coated and can be bent in any direction. The chandelier versions consist of 15 or 30 fabric-coated flexible arms, which can be shaped and aligned. This enables the appearance and lighting effect of the chandelier to be changed at will – for example wild and playful or strict and bundled.

Reef

Design Agency: Serien.Lighting
Designer: Next space

The unusual material used for the shade—ceramic foam—gives the Reef family its inimitable character. With its corallike texture the material has unique light characteristics and extraordinary design qualities: switched on, the shade shimmers in gentle, atmospheric light that is diffused in a subdued manner throughout. From the opening in the shade, however, light is emitted at an extremely wide, bright, shadow-free angle. This enables either an entire room to be illuminated from the ceiling or alternatively a small area to be highlighted on its own. Reef Wall and Ceiling for example are ideal for corridors, bathrooms and stairwells. Thanks to its low thickness, a mere 95 mm, the ceiling luminaire is particularly suitable for low ceilings. Reef Suspension has its name written all over long tables, while Reef single lends itself to small bistro tables and, lined up in a row, counters. And Reef Table shines its pleasant light on sideboards and bedside tables.

Annex

Design Agency: Serien.Lighting
Designer: Uwe Fischer

Like the archetype of a luminaire category the pure, geometric body stands in a case of mouth-blown crystal glass. The shape of the case corresponds with the exhibit it displays to create a formal whole with it. Simultaneously, it assumes the function of shade and is attached to the ceiling fixture – without distracting details. Annex is available in three sizes and several versions, so that various light effects can be achieved for high and low rooms. The light can be made to appear more dramatic by inserting an aluminum reflector; this produces direct lighting downwards and soft light on the ceiling. And a third version with a shade of frosted case glass but without an internal reflector Annex lends itself to even lighting.

Eban

The Eban brand is born in 1970 and specialises in solid wood bathroom furniture. Eban style distinguishes itself for the uniqueness of its pieces, the research of construction solutions as well as owing to the quality of raw materials used.

Eban means massive wood and skilled competent woodworkers. This vocation evolved from a traditional family experience to a modern industrial production; the up-to-date technology walks together with an antique learning and sensibility in every step of the production, until you have a high quality furniture line using environmentally friendly production methods.

E-MOTION MORGANA

Designer: Eban
Dimensions: 106cm x 51cm x 60cm

Eban creates bespoke bathroom furniture, for the customer who values quality and exclusivity. This unique experience in an intimate and warm environment will restore your sense of inner harmony

MIGNON AMBRA

Designer: Eban
Dimensions: 120cm x 36cm/51cm x 60cm

Eban offers a wide range of finishes including gold and silver versions, in addition to the choice between several handles and multiple solutions for mirrors, even with incorporated neon light, ceramic or crystal tops with different shades.

Jarrod Lim Design Studio

Jarrod Lim Design Studio focuses on developing creative concepts for furniture and products as well as generating exciting spatial designs for interiors, exhibitions and public locations. The multidisciplinary studio also assists companies with brand management, art direction and product development.

Jarrod's designs focus on the human aspects often found in everyday situations and locations. His style is intuitive and refined but always unique, with an uplifting presence. Each piece or space has its own captivating story to tell, often found hidden in the smallest details of the design.

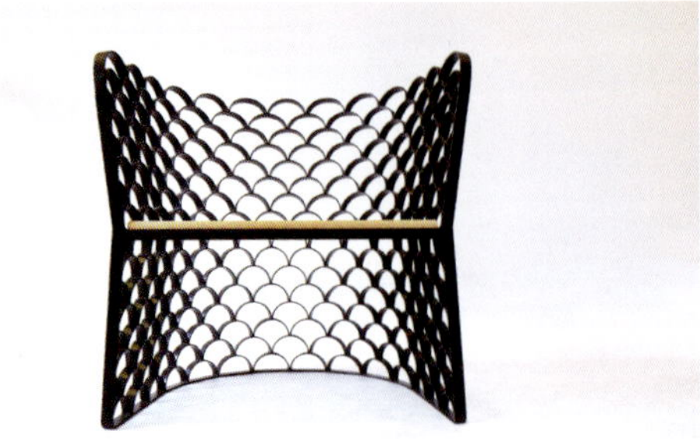

Koi Chair

Design Agency: Jarrod Lim Design Studio

The unique pattern of the Koi Chair is inspired by many of the urban surroundings found in Singapore, most notably the wrought iron gates and window grills seen everywhere on local houses and apartments.

The concept of using a gate or fence to surround a garden influenced the idea of how an outdoor garden seat can act in much the same way—as a personalised fence, but in a much more pleasant and attractive fashion.

Idle Chairs

Design Agency: Jarrod Lim Design Studio

The Idle Chairs evoke memories of times spent relaxing comfortably, enjoying a moment of peace and tranquillity.

Trinity series

Design Agency: Jarrod Lim Design Studio

The Trinity series is influenced by antique European pedestal tables combined with a modern, casual dining experience. The wooden collection employs the strong skills of local Southeast Asian craftsmen to develop products, which exemplify the mixture of Eastern and Western influence.

Collar magazine rack

Design Agency: Jarrod Lim Design Studio

The Collar Magazine Rack has two areas for storage. The upper area holds several magazines while the lower inverted V-shape can save the page in your current magazine.

2creativo Design Team

2creativo is a dynamic team with over 10 years experience in the design business working from the 22@, the innovation district of Barcelona. They help companies connecting with their customers, by defining a solid strategy and designing both their products and their brands. The priority in their projects is to simplify the life of users through simplicity and common sense. They are a strategic support for companies launching new products: they develop innovative projects to successfully introduce products or services in the market. 2creativo's work has been granted with several international design awards and has attracted significant media attention. Since year 2009, 2creativo works for companies from Hong Kong China and Southern China (Shenzhen, Guangzhou, Dongguan ...) helping them with the design of home products such as furniture and housewares following the western tastes.

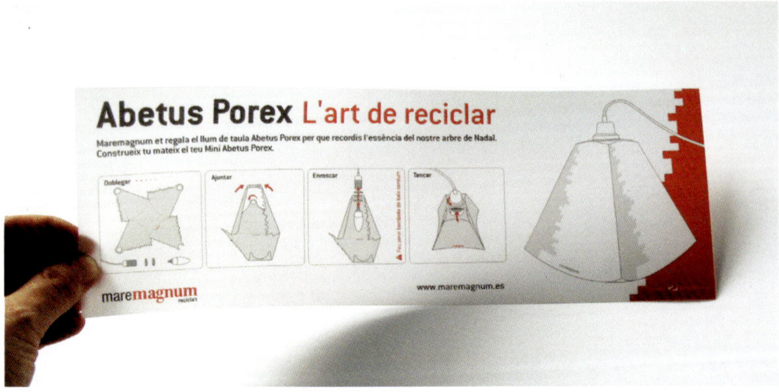

MINIABETUS LAMP

Design Agency: 2creativo Design Team

MiniAbetus table lamp is a lamp made out of one flat shade piece that gets its shape by getting folded. It has clean and minimal look. When it's on, subtle shadows are casted by the shade flaps.

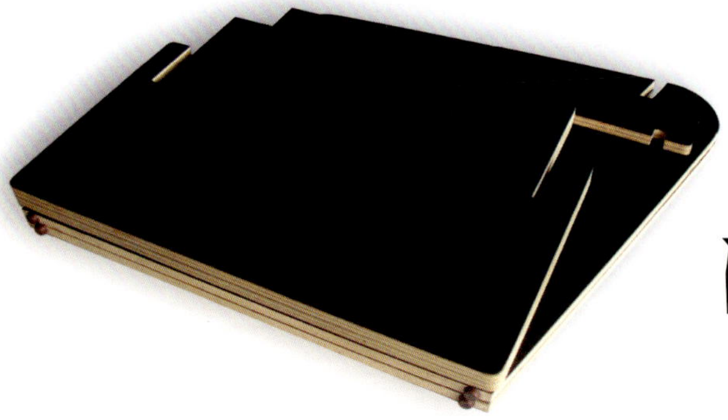

FLAT KIT ARMCHAIR

Design Agency: 2creativo Design Team

Flat Kit is a collapsible armchair made out of plywood board without using glue, nails or screws. It is elegant and very comfortable and can be assembled and disassembled easily converting into a lightweight flat pack.

Christina Waterson

Australian artist, designer and architect Christina Waterson has a vibrant creative practice making artworks, realizing large-scale public art projects, designing exhibitions and events, and developing innovative product ranges for manufacture. Waterson is inspired from natural geometries and growth patterns as well as the properties of the materials she uses. Her considered variations in scale, materiality and combination have generated beautiful and distinctly different objects, product ranges and environments to enjoy. One can feel Waterson's special care and attention through the pleasure of directly experiencing her works.

THE BLOOM SERIES

Designer: Christina Waterson
Materials: refer to details of each design and include certified plywoods (seat, mag rack, lights and platter), recycled corflute (screen), recycled perspex (ornaments), stainless steel (bowl)
Photography: Jon Linkins

The Bloom Series is a seven-piece furniture and homewares range that celebrates natural geometries and growth patterns. It includes a bowl, platter, seat, magazine rack, lights, screen and ornaments; each combining male and female pairings and using a range of materials to make a series of inspirational objects.

H220430

Design & Produce: Satoshi Itasaka
Design & Computer Graphics: Takuto Usami
Satoshi Itasaka and Takuto Usami collaborated to establish 'H220430' on April 30th, Heisei 22(Heisei is the current era name in Japan). Their design activities focus on lighting and furniture, etc.

LARSEN C

Design Agency: H220430
Dimensions: 9cm x 9cm x 8cm

In the Antarctic Peninsula, there is one of the largest ice shelves called Larsen. It consisted of a series of three shelves, namely Larsen A, B and C, from the north. Ice shelf Larsen A was regarded to have existed for more than two thousand years before its disintegration in 1995. Then, in 2002, huge ice shelf Larsen B broke up after an existence extending back 12 millennia. Those disintegrations in recent years caused a tremendous amount of ice loss, which corresponds to an area more than six times as large as Tokyo. Only ice shelf Larsen C remains, but unfortunately, its disintegration, too, is just a matter of time.
H220430 created an acrylic object and named it Larsen C, which will remind you of an iceberg split from a disintegrating ice shelf floating in the sea. It has two holes in the top, and you can use it as a toothbrush holder or, by putting it into another vessel with water, as a bud vase. When it is wet, Larsen C looks even more like a real iceberg melting off the shelf, which will make you speculate more realistically about the serious condition of the Antarctic area, even though it is far away. H220430 would be glad if the users come to think about global warming and consequently take some actions to prevent it by using Larsen C in their daily lives.

Rubber stool

Design Agency: H220430
Dimensions: 32cm x 32cm x 40cm

Regarding the rubber, which has widely used for a long time, a research and development of rubber is moving forward with a progress of the material science. Now, the artificial synthetic rubber becomes popular and plays an active part in this field.
On the other hand, a demand of natural rubber keeps expanding since natural rubber is better than synthetic rubber in terms of its physicality and cost. This enlarges a large scale of rubber tree planting, and environmental destruction due to a deforestation become worse mainly in Southeast Asia. Although to use recycled rubber is proposed as one of the approaches to improve this circumstance, this approach is not progressed. H220430 designed a stool made by recycled rubber taking this situation into consideration.
Although a framework of this stool is simple with just bending one piece of rubber plate and holding legs with bolts, a form of this stool is unimaginably elegant and comfortably cushioned with elasticity of rubber. This stool can be stored in a small space by rolling up when it is not in use.
H220430 hope that this Rubber Stool is widely used as one of the usage of recycled rubber and it would be a trigger that people realize the circumstance of deforestation for making rubber.

Porada

Porada rises up from the furniture field thanks to refinedness, simplicity and essentiality of the lines, to timeless design, attention for details and handmade tradition-oriented workmanship. The company Porada was established in 1968 by Luigi Allievi with the help of his 4 sons in order to expand the production of chairs, started in 1948, to other branches of the home furnishing. Porada is based in the economical contest of Brianza, famous for the industrial district of wood furniture, defined by an excellent execution of the different phases of product making. Nowadays, Porada cooperates with several architects such as Mr. Colzani, Marconato & Zappa, Seveso & Trezzi, Gottein, Gonser and Studio Opera Work. Porada is the excellent combination of technology with the ancient local craftsmanship, its collection is at the same time classical and contemporary creating 'classical, modern and timeless' products. Porada joins solid wood to other materials, as glass, steel and metal, paying attention to details, and mixed all these features in order to create a wide collection of chairs, tables, mirrors, coffee tables, TV trolley and so on.

Infinity

Design Agency: Porada
Designer: Stefano Bigi

Low table with glass top and solid wood base, fixed on a chromed metal ring.

Nissa

Design Agency: Porada
Designer: Marconato & Zappa

Chair with solid wood frame and cover in the fabrics of the collection.

Alba

Design Agency: Porada
Designer: M. Walraven

Armchair with frame in solid ash or canaletto walnut and removable cover in the fabrics of the collection.

Rapsel

From the very beginning in 1975, Rapsel carried forward its creative belief with consistency and determination. The company's policy is to operate in the area of contemporary design, identifying objects with highly innovative content and great quality of aesthetics and form. Research is the basis of all their projects.

Rapsel's trend-setting objects have great character and in some cases they have even been described as 'cult objects'. However, these are not only objects of mere appearance, but they are functional and from the technical point of view perfectly made.

ELLE

Design Agency: Rapsel

CHEF

Designer: Marco Merendi for Rapsel

Chef reminds of the delicious kitchen world: made of aluminium in different colors and sizes, these basins resemble huge cooking pots and will engage the young at heart. The basin freely stands on an oak wood table or a pedestal. A poetic mirror, a container and a soap holder as well as a toothbrush glass holder complete this innovative new series.

OFURÒ

Designer: Matteo Thun and Antonio Rodriguez for Rapsel

Inspired by the Japanese bath that serves in particular as a treat to relax and recuperate from the daily routine, Matteo Thun has created a new bathtub with a European touch: OFURÒ. Its puristic form allows complete emersion, while its material and composition convey calm and warmth. It is produced entirely from larch wood which is dried in several work steps and cut, shaped and joined together in a special procedure. OFURÒ is haptic, visual and comfortable: one of a kind.

Dario Gagliardini

Dario Gagliardini was born in Milan. He is doctorate on architecture by the Technical University of the same city. He's collegiate architect, condition that allows him to live through some significant experiences in the works of the hotel company CIGA and of the Catterpillar Italy. He designs products and to the communication of the same ones as visual designer and art director, extending with some companies his contribution like design management. At the same time, he is journalist and collaborates, like graphic designer and art director, with prestigious design and architecture magazines like Ottagono and all ones published by Design Diffusion Edizioni (DDN, OFX, Design Management, Casa D, etc).

Arko

Designer: Dario Gagliardini

Classic topics with contemporary essence. Designed and coordinated by Dario Gagliardini. The Arko collection is introduced like a reinterpretation of the classic topics with a contemporary essence. The combination between modernity and tradition turns Arko into the ideal product to decorate the interiors of sophisticated style. The modernity of its lines evokes traditional pieces of contemporary taste.

DIBA

Designer: Dario Gagliardini

The model Diba has two proposals: an armchair and an armchair relax. The aim characteristic is the two versions of the back: high and low. Contemporary armchairs with character. Stylized and comfortable lines. The soft shapes are due to the round curve of the armrest accentuated by the exclusive legs of stainless steel.

Sander Mulder

Sander Mulder's work is proof of the firm believe that good concepts can only exist in symbiosis with a good execution, and it investigates new ways of approaching the construction, functionality and aesthetics of everyday objects and spaces. Supported by the believe there's no constant except change and that joy is the accelerator of creativity, he's not afraid to wander aimlessly and conduct experiments in search of the coincidental beauty of failure. The resulting work is a combination of strong lines, subtle details, innovative techniques and bold statements, often questioning archetypes and their respective interpretations by mankind.

Marie-louise

Designer: Sander Mulder
Dimensions: 60cm x 38cm
Materials: acrylic, polycarbonate
Photography: Niels van Veen & Sander Mulder

Made from 16 transparent contours illuminated by a fluorescent bulb, this table lamp provides a modern reincarnation of the traditional ambient lamp. Through the use of special materials and accurate CNC milling techniques, not only the hood section, but the whole body of this fixture emits a magic light.

CARAT

Designer: Sander Mulder
Dimensions: 46cm x 66cm
Material: anodised aluminum
Photography: Frank Gielen & Sander Mulder

Diamond cutting requires art, skill and, increasingly, science to transform a rough stone into a faceted gem. This intricate modular lampshade design was conceived with the same ingredients. It consists of 9 identical parts and an astounding 54 facets. The anodized aluminum finish on the outside of the shade makes it gleam like an emerald in the dark, while the reflective inside allows for optimal lighting.

Ceci

Designer: Sander Mulder
Dimensions: 11cm x 6cm.
Materials: pyrex bulb, high power LED, transformer

This lamp plays with the stereotypes of our daily lives which are imprinted in our minds. By clever use of a high power LED, lamp becomes shade, and the impossible becomes possible. The bulb is hand-blown from heat-resistant Pyrex glass, and therefore without sharp edges also safe to touch. The used power LED is an environmental friendly and virtually indestructible light source compared to an average incandescent bulb.

Pandora

Designer: Sander Mulder
Dimensions : Small: 50cm x 100cm x 50cm/ Medium: 100cm x 100cm x 50cm/ Large: 200cm x100cm x 50cm
Materials: powder-coated steel, stained glass
Photography: Niels van Veen & Sander Mulder

This modular storage system is inspired by the wonderful color mosaics that sprout to life in every harbor and container terminal the world around. The individual pieces can be stacked and rotated around in endless (color-) combinations, to create your personal container terminal for all domestic storage uses.

Dsignio

Dsignio is a integral design studio located in Madrid that executes international projects. It was founded by Alberto Bejerano and Patxi Cotarelo in 2002, both of them are technical engineers of industrial designs and interior design master.

Dsignio focuses its activities in the industrial and interior designs, working for national and international companies. They have received several prizes and have participated in different showrooms. Its projects have been published in Experimenta, Proyecto Contract, Diseñart, Nuevo Estilo, Habitania, Woman, Marie Claire, Horeco, Equipamiento Hostelero, Oxígeno, ABC, Telemadrid, etc.

The professional practice has been combined with teaching, taking part in conferences and classes in different universities in Spain and Mexico.

Link

Designer: Dsignio

We try to design a timeless sofa with a simple geometry, which could be adapted to the contemporary architecture. It was necessary to modulate it in order to create different compositions with ergonomics and well care for comfort.

GSG Ceramic Design

GSG is a company specialized in the production of ceramic sanitaryware, and is to be found in the ceramic district of Civita Castellana. The GSG company has a long experience in the foreign market with classic design, and now, a careful awareness of today's designs and styles, enlarging the gamma of products to combine a new way of living and concept of architectural bathrooms.
From this idea the GSG Ceramic Design has realized an esthetic result, functional and modern, developing the Race, Lilac and Dunia series—three projects for one bathroom design.
The 2006 collection is completed by 'oz' washbasin and the shower tray 'ecoscape'—two realizations that go beyond a simple sanitaryware project.

Foto Cat Colore City LR

Design Agency: GSG Ceramic Design

GLASS SYSTEM

Design Agency: GSG Ceramic Design

Lightness, innovation, consistency, cleanliness, are prerogatives contained within the new concept of the GSG Ceramic Design.
The extreme formal linearity of glass, creates a shell for water connecting the ceramic forms which support the elements in cristal. It is a strong architectural concept : emptiness and transparent overlappings conjointly with supporting ceramic elements, giving one the sensation of intangibility. Cristal evokes faint reflections, always indefinite; ceramic supports the power of gravity creating detectable limits.

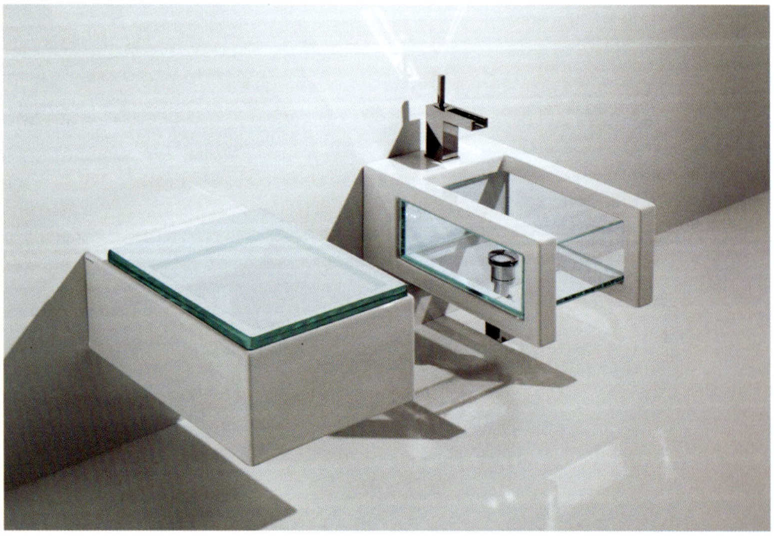

FOTO CAT COMPL

Design Agency: GSG Ceramic Design

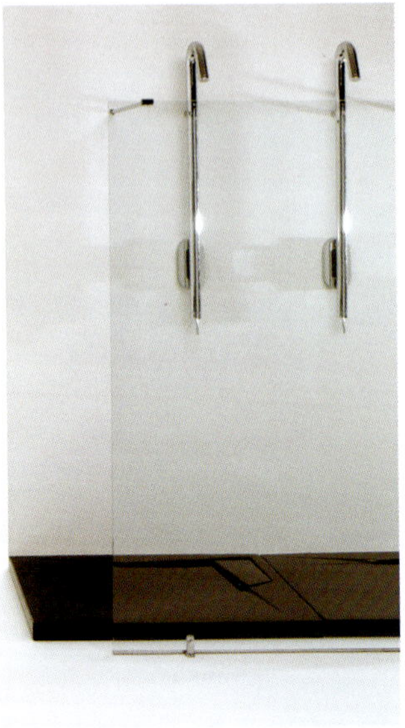

Pallucco

A natural inclination to continuous research, on both lighting and living sectors, that, just like the pieces of a jigsaw, come together to create a range of design ideas for contemporary living that has international appeal. Since 1984, year of establishment of the brand, Pallucco drove its way to innovation and dialogue with the world of design.Reissuing and creation, rediscovering and experimentation , are the principles on which moves the vision of Pallucco company. The attitude to rediscover projects of the past live together with frantic curiosity to see first the drawing of a young designer. Visions of masters of design cohabit with hazard of the younger, that in this way can try out his talent suggesting an iconic design for Pallucco.

Foglie

Design Agency: Pallucco
Designer: Matali Crasset

The pendant versions are available in three different diameters: 520mm, 680mm and 800mm. The diffusers on the floor lamps can be: 520 mm diameter and 1800mm height or 680mm diameter and 1820mm height.
Each lamp is fitted with a Globo 30 Watt Ø120 compact fluorescent light bulb E27 fitting (220V). 23 Watt E26 fitting for the 110V version. Warm light 2700° K

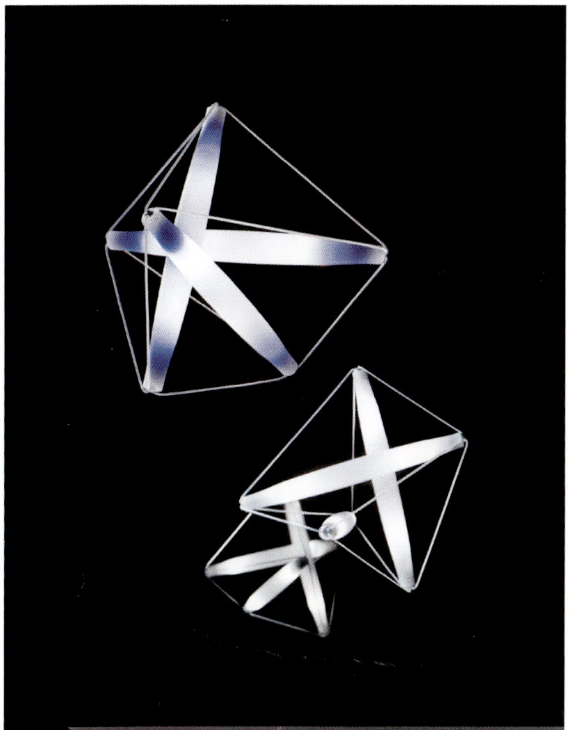

Buck

Design Agency: Lagranja Studio for Pallucco

The lamp is fitted with six energy saving 40 W halogen light bulbs (two for each rod) with a G9 socket (220V and 110V).

Tabard

Design Agency: Pallucco
Designer: Denis Santachiara
Dimensions: 60cm x 67.5 cm x 42cm.

Fitted with one 20 Watt fluorescent electronic light bulb Ø100, socket E27 (220V) and 16 Watt, E26 fitting for the 110V version.

Constance Guisset

- Creator of the year, Maison et Objets 2010
- Stage Design for Laurent Garnier's concert at Pleyel concert hall, Paris
- Conception of Galeries Lafayette Maison windows for Designers'days 2010
- Janus de l'industrie award for Xtremkey USB key design for LaCie
- Scenography for 'Petites et Grosses Bêtes' exhibition at Musée des Arts Décoratifs, Paris
- Edition of Vertigo by Petite Friture
- Winner of Audi Talents Awards 2010
- Exhibition at Spazio Rossana Orlandi, Milan, Milan Furniture Fair 2010
- Elle Decor, Young talents on show, Superstudio Piu, Milan 2010, collective exhibition

Dancing chair

Designer: Constance Guisset

Dancing Chair — Aide à projets, VIA 2009, presented in Paris and Milan, Furniture Fair 2009 — is a rocking chair and a footrest ensemble. The seat rests in the wooden structure, like suspended in a crinoline.

Tri3

Designer: Constance Guisset

Tri3 is a garbage can of sorting. The left pedal opens the highest can in a traditional way. This can is for general discarding. Bag dispenser. The middle pedal makes the second can rotate around a decentered axe. This can is for packaging, plastic, etc. The right pedal makes the last can move towards the user. The last can is for glass discarding.

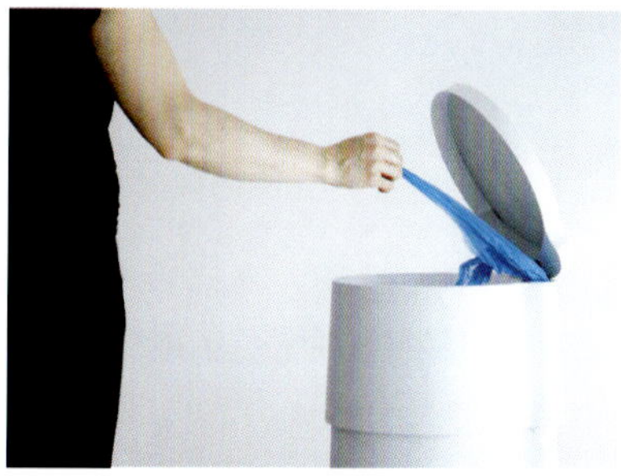

Studio Gorm

Studio Gorm is a collaboration between John Arndt and Wonhee Jeong. The two met while studying in the master's program at the Design Academy Eindhoven and shortly after they set up their studio in Rotterdam, the Netherlands. Their office is now located in Eugene OR where they also work in the new Product Design Program at the University of Oregon. The two come at design from different approaches to produce smart objects without pretension and warm comfortable environments. Their products are simple, practical and thoughtfully made. Since their debut at the Salon de Mobile in Milan in 2007 their work has been featured in numerous international publications and exhibitions.

Nr 3 chair

Design Agency: Studio Gorm

This strong simple chair draws its influences from two classic pieces of Egyptian furniture. The seat is laminated in large sections, which can produce a bench or two chairs. The triangulated joint makes it extremely strong. This is a chair you can pass on to your grandchildren. The fabric version has a comfortable fabric sling seat, the chair can be easily dissembled for storing and shipping.

Nr 3 fabric chair

Design Agency: Studio Gorm

FLOW 2 KITCHEN

Design Agency: Studio Gorm

The flow products can be used independently but are far more effective when they work in concert as part of a larger system. The individual objects are relatively uncomplicated, acting as simple vehicles for the more complex natural processes to do the work. This kitchen is developed as a flexible system where resources are reused by several elements creating a dynamic flow between the products. The flow kitchen focuses on three major areas: Waste, Water and Energy.

POSTFOSSIL

POSTFOSSIL is a collective of emerging designers, which creates objects for the home, thinking about the postfossil reality and putting their objects into the context of resource scarcity. The use of environmentally friendly and renewable resources is only the basis; POSTFOSSIL takes it to the next level: fossil does not only stand for energy sources but also for social behaviour patterns. That's where the initiators start.

The collections include products that focus on the use of materials; others attempt to reflect everyday – fossil – routines, to reconsider habitual processes. Others deal with the subject ironically. The collections from POSTFOSSIL aspire to long life and environmental friendliness, and reflect a responsible use of resources. POSTFOSSIL does not want to limit itself; this variety of interpretations offers space, and this is one of the collection's strengths. The POSTFOSSIL platform wants to design products for the postfossil age and offers services, which will encourage discussion about resources, consumption, and society.

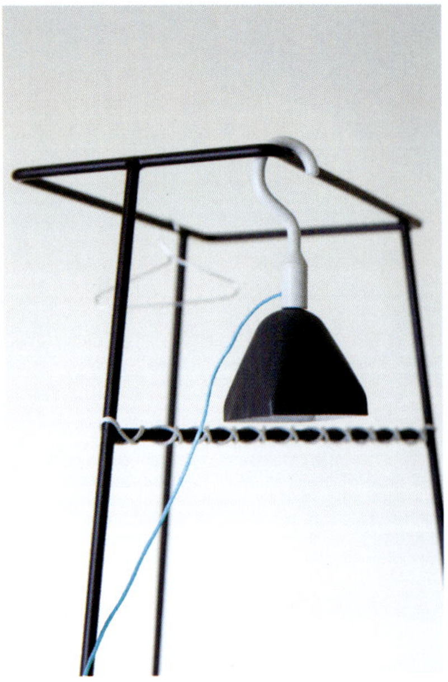

ACTIVE LIVING, HOMELIKE WORKING 2010

Design Agency: POSTFOSSIL

To combine old with new, expensive with cheap is nowadays custom. The different requirements for living and working today ask for adaptable and changeable design. With 'active living, homelike working 2010' the idea of a showroom is picked up, developed further and designed for the individual needs today. It comes out a fusion of working and living furniture, which allow for individuality and sustainability.
With 'active living, homelike working 2010' POSTFOSSIL won the second prize at the Werkbund design® Embru competition, November 2009.

Jason Phillips Design

Noteworthy new designs include the spaghetti-like Continuum Table, which was inspired by models of the space-time-continuum where folds in space and time are created through invisible networks. Al Dente is a decidedly different take on fluidity; relying on rigid rods that explore a more fractal sense of space and density. These concepts are in production and will be introduced later this year. Most recently, Jason was awarded the inaugural INSPIREDESIGN Design Excellence Award. The previous month he was honored at the 21st Annual ARTS Awards as 2010 Product Designer of the Year, the youngest ever recipient of this prestigious award. In late 2009 he won the VERANDA magazine's inaugural 'The Art of Design' Award for his body of work in the 'Trade Secrets' category, which has led to a whirlwind of media attention.

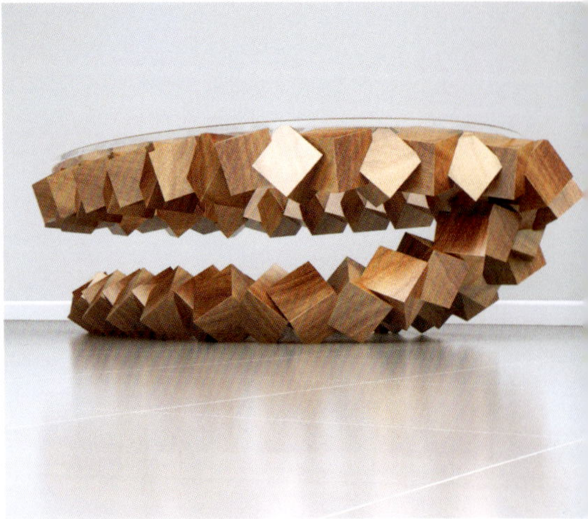

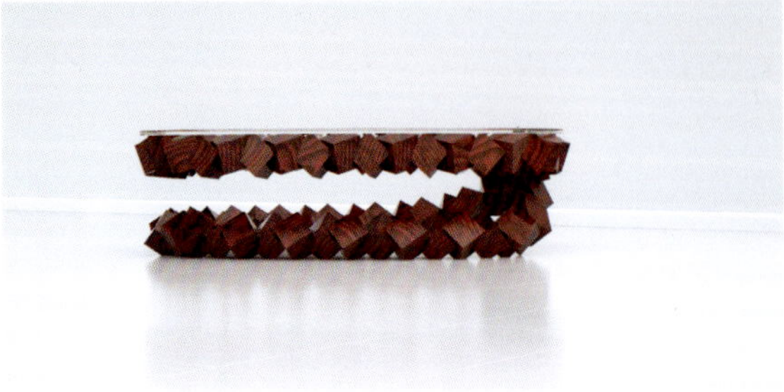

Corocotta Table

Design Agency: Jason Phillips Design

According to several historians, Corocotta was a Cantabrian warrior leader during the 1st century BC. His great achievement was the union of the disperse Cantabri clans in an alliance against the invading Roman armies, to the great exasperation of the powerful empire.

Monoliths

Design Agency: Jason Phillips Design

A monolith is a geological feature such as a mountain, consisting of a single massive stone or rock, or a single piece of rock placed as, or within, a monument. Erosion usually exposes the geological formations, which are most often made of very hard and solid metamorphic or igneous rock.

Nilufar

Nilufar is known today as one of Italy's most active and original galleries in the fields of historical and contemporary design, antique Oriental carpets and furniture. The gallery—now housed in a three-storey space on Milan's prestigious via della Spiga—was founded by Nina Yashar in 1979 and, since then, has presented exhibitions of an unmistakable style: always eclectic, and always to be read in novel, unexpected ways. Becoming an inescapable reference point to lovers of historical design as well as to people following the research and evolution of contemporary design, above all within that more learned, poetic and visionary area shifting between industry and art, product and artistic work.

Totem

Design Agency: Nilufar
Designer: Laura Bethan Wood

Pyrex light made from interlocking elements that stack together to create a totem of glass and light. An extension of my interest and research started with an entwined light made in collaboration with Seguso Glass. This light plays with the transparency and thinness of Pyrex, exploring the typical forms associated with the medical and chemical functions of Pyrex.

Chest of drawers

Design Agency: Nilufar
Designer: Martino Gamper

RITA

Design Agency: Nilufar
Designer: Martino Gamper

ZAAFDesign

ZAAFDesign is an industrial design studio from Torino, Italy, which breathes the air of design and creativity of various latitudes and longitudes. It has designed and still designs products for companies based in Europe, Middle East and Asia. When talking about itself, ZAAFDesign's team likes to use 3 words: project, creativity, innovation. The capacity of vision, the sensibility of catching new signals and knowing how to transform ideas into realizable proposals allow them to develop products and services which are in line with corporate strategies.

Sinuosa

Designer: Federico Rossi, ZAAFDesign
Manufacturer: Artie Garden International, Ltd.
Material: synthetic rattan

Conceived from a graphic sign sketched on paper, a very light and sinuous chaise lounge was created, through light bending and controlled thickness of the material.

SASSO

Designer: Federico Rossi, ZAAFDesign
Manufacturer: Artie Garden International, Ltd.
Material: synthetic rattan

SASSO is conceived from a carved rock. Through the emptying process a single and a double seat have been created on opposite sides. Made of synthetic rattan the weaving is done around a complex tubular structure.

BREZZA

Designers: Federico Rossi and Giovanni Duro, ZAAFDesign
Manufacturer: Wisdom Sanitary Ware Co., Ltd.
Dimensions: 237.1cm x 193.4cm x 41cm
Material: solid surface

The Brezza Bathtub concept permits to change the classical bathtub through the passage from tradition to innovation, changing an individual object into a convivial and multi-functional one: the side is reduced to one layer, that seems shifted by the wind, expressing the idea of lightness. The bathroom becomes the place for the social wellness and gathering.

Diego Granese

Diego Granese, Italian architect and designer, who lives and works in Salerno. In 2000 he founded the design and architecture studio Granese, an interdisciplinary workshop which provides projects services to a selected clientele. In 2003, together with the Spanish company Frajumar, he won in the professional category of 'XXI International Contest of Design CDIM' in Valence, Spain. In 2004, he was selected by ARCO International Contemporary Art Fair in Madrid, in the Index ADI, in BIO 19 and received the recognition in the XX Compasso d'Oro.
His products have participated in several transmissions in television and Spanish movies. His products are part of a lot of collections like 'historical collection of Compasso d'Oro' and permanent collection of the Museum 'Pinacoteca of Modern Art' of Munich in Baviera, Germany. Nowadays, he executes several projects of architecture and interior design in Italy and abroad. Among others, he produces the limited edition distributed around the world.

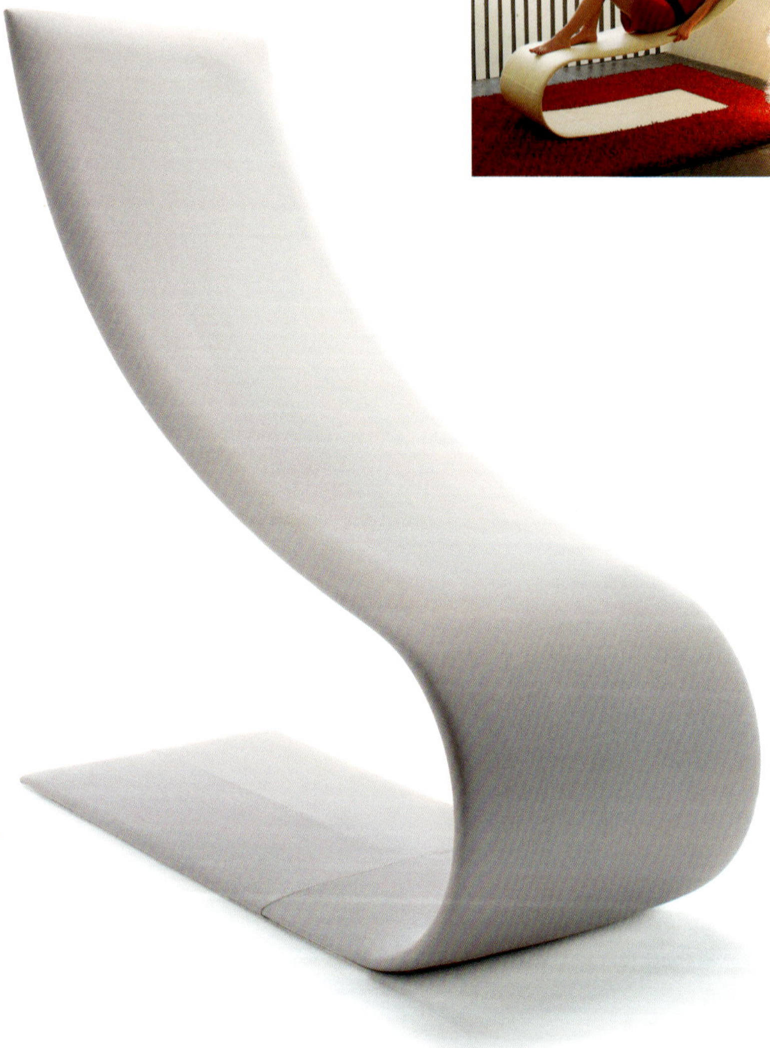

ONDA

Designer: Diego Granese

A new piece of furniture of synthetic and minimalist concept that varies and transforms into a seat made of a single piece of steel. The maximum resistance estimated to deformation is from 200 kg, although the extension varies according to the sitting and the place where the extension force is applied. The weight of this design has been calculated in 100 kg, but it has a direct connection with the safety of the product depending on the function and use given to this piece. The weight is a resultant of shape, material, adaptation of human figure and mechanism. This product has been recognised and awarded by the furniture sector.

CUCHARA

Designer: Diego Granese

Cuchara is an armchair/ rocking chair for interiors, made in built-in polyurethane, with a steel base and leg. It is obvious that the characteristics of the materials offer a variety of colors, to be suitable to all environments: from the bedroom to the living room, the relax area… but also in a desk room and for television programs. Designed by Diego Granese.

Zanotta

Established in 1954, Zanotta is one of the recognized leaders in Italian industrial design. Today's collection comprises these furniture types: armchairs, sofas, beds, storage units, bookcases, chairs, tables, coffee tables and furnishing accessories. Diverse manufacturing technologies are employed. In harmony with the corporate philosophy aimed at R&D, they allow the advanced use of a host of materials, like metals (aluminum alloy, stainless steel, brass, bronze, etc.), plastics, glass, marble, granite, wood, fabrics and leather.

CALLA

Design Agency: Zanotta

Based on castors or fixed with feet. Steel frame. Upholstery with flame-retardant polyurethane foam/heat-bound polyester fibre with elastic strips suspension. Removable fabric or leather cover.The external body can even be covered with cowhide combined with fabric or leather. The armchair is available in two different heights above ground level: high 873/A with castors, low 873/B with feet.

Loemen

Mireille Verroken & Dries Braeckman are the Belgian light designers behind the Loemen light sculptures. The creation of a Loemen light sculpture starts from an idea, a visualisation of a shape, that is then turned into a handmade mould. In this way an abstract concept is materialised into a concrete form. The mould functions as the basis for the final light sculpture that is made out of plastics whether or not clothed with glass fraction, which is a unique process. The design as well as the creation of the light sculptures is entirely done by themselves. The Loemen light jewels are 100% handmade in their atelier in Ronse (Belgium), the creative center of their artistic activities.

Their mission is to fuse magnificent light jewels into exquisite architectural environments—both historical or modern—in order to create oases of beauty based on precious handmade and soul-changed pieces.

THE PINK SATELLITE

Design Agency: Loemen

THE SATELLITE AND MYCENA LUX

Design Agency: Loemen

The Satellite and Mycena Lux are 2 out-of-the-box Loemen light objects that can either be installed against a wall or against a ceiling. They are made of artificial material combined with brilliant glass fractions. Color and pattern can be chosen.

JUSTIME Design Team

JUSTIME Design Team, in house design of SHENGTAI BRASSWARE CO., LTD (established in 1976), was established in 1998. The membership is 10 designers who are in charge of different domains from first to last, such as appearance, structure and package design. JUSTIME Design Team is the proficient on Faucet & Shower, Bathroom Accessory & Furniture design. Consequently, JUSTIME Design Team does have a great achievement on worldwide stages of design awards, for instance, red dot product design award, Good Design award (Chicago), iF product design award, KAPOK Prize—China Innovative Design Award (Best of The Best).

STILL ONE

Design Agency: JUSTIME Design Team

Inspired by main concept of 'Less is More', Still One uses least lines to integrate practicality and elegance into the concise design. Main material is lead-free brass or stainless steel and meets NSF standards to ensure drinking safety. The valve, made of stainless steel, extends lifespan and can be used easily.

Story series

Design Agency: JUSTIME Design Team

JUSTIME hopes the design could remind adults of past about original charming and also make kids experience the beauty of traditional and modern times. Story Series tells about history of faucet: from past, through present, and to the future. As people talking about faucet, the first impression is 'Story Faucet'.

Tai chi series

Design Agency: JUSTIME Design Team

Tai Chi series was inspired by the sculptor Master Zhu Ming who presented the beauty of the motion of human body as the work of art and the concept of Tai Chi, JUSTIME Design Team utilizes its abstract perspective in this designing of Tai Chi Series faucets.

Acknowledgements

We would like to thank all the designers and companies who made a significant contribution to the compilation of this book. Without them, this project would not have been possible. We would also like to thank many other people whose names do not appear in the credits, but who provided valuable input and support for the project from the start to its completion.

Future Editions

If you would like to contribute to future Artpower publications, please email your details to: info@artpower.com.cn.